Barley, Malt and Ale in the Neolithic

Merryn Dineley

BAR International Series 1213
2004

Published in 2019 by
BAR Publishing, Oxford

BAR International Series 1213

Barley, Malt and Ale in the Neolithic

© Merryn Dineley and the Publisher 2004

ISBN 9781841713526 paperback
ISBN 9781407326245 e-book

DOI https://doi.org/10.30861/9781841713526

A catalogue record for this book is available from the British Library

This book is available at www.barpublishing.com

BAR Publishing is the trading name of British Archaeological Reports (Oxford) Ltd.
British Archaeological Reports was first incorporated in 1974 to publish the BAR
Series, International and British. In 1992 Hadrian Books Ltd became part of the BAR
group. This volume was originally published by John and Erica Hedges in conjunction
with British Archaeological Reports (Oxford) Ltd / Hadrian Books Ltd, the Series
principal publisher, in 2004. This present volume is published by BAR Publishing,
2019.

BAR
PUBLISHING

BAR titles are available from:

BAR Publishing
122 Banbury Rd, Oxford, OX2 7BP, UK
EMAIL   info@barpublishing.com
PHONE   +44 (0)1865 310431
FAX   +44 (0)1865 316916
www.barpublishing.com

Contents

Contents..i

Abstract, acknowledgements..iii

List of illustrations...v

INTRODUCTION..vii

CHAPTER ONE: The Conversion of Grain into Malts and Ale.............................1
1. The techniques and biochemistry of brewing 2
    a) Malting: the controlled germination of the grain 2
    b) Kilning the malt...3
    c) Mashing: the conversion of starches into malt sugars...................................4
    d) Obtaining a sweet wort for fermentation 6
    e) Boiling the wort with flavourings and preservatives......................................9
    f) Fermentation; the conversion of sugars into alcohol 9
2. Necessary materials and equipment 10

CHAPTER TWO: Additives to Preserve and Flavour Ale 13
1. Additives used in Medieval and Viking times..13
2. Archaeological evidence: herbal additives in prehistoric Europe.......................14

CHAPTER THREE: Barley in the Levant, Ancient Near East and Egypt.............19
1. The 'bread/beer' debate 19
2. Hunting and gathering groups: Near East/Levant in the 9th and 8th millennia BC ...19
3. Early grain cultivation and processing in the 7th and 6th millennia BC.............22
4. Early pottery neolithic cultures in the 6th and 5th millennia BC......................23
5. Fermentation of barley wort in the 4th and 3rd millennia BC 25
6. Brewing in Egypt in the 3rd and 2nd millennia BC..26

CHAPTER FOUR: Grain in Neolithic Europe 29
1. The transition to agriculture in Europe...29
2. European Neolithic: 6th and 5th millennia BC..29
3. Late Mesolithic and early Neolithic: 5th and 4th millennia.............................33
4. Neolithic groups: 4th and 3rd millennia BC 36

CHAPTER FIVE: The Stone Buildings of the Orcadian Neolithic 41
1. Neolithic settlements of Orkney: 4th and 3rd millennia BC.............................41
    Knap of Howar, Papa Westray 42
    Barnhouse, mainland Orkney...48
    Skara Brae, Skaill Bay, mainland Orkney...51
    Rinyo, Bigland Round, Rinyo 58
    Other Orcadian settlements 60
2. Women in Neolithic Orkney 60

CHAPTER SIX: The Grooved Ware Culture in Neolithic Britain.........................63
1. The transition to agriculture 63
2. Suitable pottery vessels for ale 63
3. Organic residues 65
4. Suitable buildings 68
5. Durrington Walls 70
6. Interpreting Neolithic timber buildings 71
7. The Grooved Ware culture 72

SUMMARY AND DISCUSSION 75

CONCLUSIONS 77

BIBLIOGRAPHY 79

ABSTRACT

The keystone of this thesis is that the biochemical laws that govern the processes of malting, mashing and fermentation remain unchanged throughout the millennia. Therefore, the practical, science-based experiments are a valid means of examining these processes and of assessing an assemblage pattern or 'archaeological signature' for grain processing, malting and brewing activity in the past.

This research began by studying the probable techniques and methods of maltsters and brewers in the Bronze Age. The focus soon turned to the Neolithic, given the identification of cereal-based residues mixed with pollen on sherds from large Grooved Ware vessels that were dated to the early 4th millennium BC, found at a ritual and ceremonial site at Balfarg, Fife, Scotland.

Barley, as well as being a source of carbohydrate in the diet is also a potential source of malt, malt sugars and ale if processed correctly. The biochemical and practical aspects of processing grain into sweet malts and ale on a domestic scale are described in Chapter One, with illustrations of the reconstructive experimental work undertaken in mashing, making barley 'cakes' or *bappir*, obtaining the sweet wort from the mash and fermenting.

If ale is to keep well then herbal additives must be added to the wort during the boil. Hops are currently used but in Medieval times and earlier, meadowsweet and other common herbs were used to preserve, enhance or flavour the ale. This aspect of brewing is investigated in Chapter Two.

Chapter Three considers the earliest grain gatherers and processors of the 9th millennium BC and the consequent development of domestication, cultivation and processing of grain throughout the Levant, the Near East and Egypt.

Chapter Four evaluates some of the archaeological evidence for these processes in Europe, from the 6th to the 3rd millennium BC. Specific cultural groups are considered, for example, the Bulgarian settlement tells of the 6th and 5th millennia BC, the widespread agricultural groups of the 5th and 4th millennia BC, known as the Linearbandkeramik and the TRB or Funnel Beaker culture of the 4th and 3rd millennia BC. The coastal European Mesolithic cultures of the 5th and 4th millennia BC, known as the Ertebølle and Ellerbeck cultures are also considered.

Chapter Five is an analysis of the well preserved stone buildings of the Orcadian Neolithic that date to the 4th and 3rd millennia BC and Chapter Six assesses the Grooved Ware culture of mainland Britain and the potential for transforming grain into sweet malts and ale. The role of women as grain cultivators and processors is also considered.

ACKNOWLEDGEMENTS

I would like to acknowledge a number of people who have been of assistance in the production of this thesis. Graham Dineley, a craft brewer of some twenty years' experience, demonstrated and explained the practical application of biochemical theory in domestic mashing and brewing practice.

The advice and comments of Dr David Coombs, my supervisor at Manchester University, were greatly appreciated. Dr Stuart Campbell, also of Manchester University, read and commented upon Chapter Three.

Arlene Isbister, an artist from Harray, Orkney, demonstrated the use of haematite as a pigment and she also introduced me to the academic work of Dr Hilda Ellis Davidson. Thanks are due to Sam Dineley for re-formatting the thesis for publication and for his work on scanning and labelling the illustrations, maps and diagrams. Every effort has been made to contact the publishers and authors for copyright permission to reproduce their illustrations.

Harry Flett, Curator of the Corrigall Farm Museum, Harray, Orkney, grew and malted Bere barley that was used in some of the mashing and brewing experiments. He shared his experiences and his practical knowledge of traditional floor malting techniques.

Fawcett's Maltsters of Castleford, Derbyshire also provided traditionally floor malted grain for the experimental mashing work. This company is one of the last remaining traditional floor maltsters in the British Isles. I would like to thank Mr James Fawcett for his interest in this research and for sharing his in depth knowledge of the ancient craft of floor malting.

List of illustrations

1.1 Suitable equipment to ferment a barley wort 1
1.2 Grooved Ware vessels from Skara Brae, Orkney..1
1.3 Inside the grain barn, Corrigall Farm Museum, Orkney 3
1.4 Ground plan of the 200 year old barn at Corrigall Farm Museum 4
1.5 The excavated remains of the U-shaped ditch at Eberdingen-Hochdorf, north Germany................................5
1.6 Carbonised malt from Eberdingen-Hochdorf..5
1.7 Sparging the wort with a grain bag and bucket 6
1.8 Egyptian maltsters, mashers and brewers...6
1.9 Pale crushed malt mixed with abundant water in a sealed pottery bowl 7
1.10 The saccharification of the barley mash..7
1.11 Rectangular hearth based on the Orcadian Neolithic 8
1.12 Crushed malt and sweet barley mash 8
1.13 Pure sweet malt liquid in a container used to store the barley mash 8
1.14 Settling and settled wort...8
1.15 Table of the basic equipment for processing grain into malt, sugars and ale 10
1.16 Pollen analysis from beaker found at North Mains, Strathallan, Scotland 11
1.17 Pollen analysis from beaker found at Ashgrove, Fife, Scotland..11

2.1 Ground Ivy (*Glechmona hederacea*)..15
2.2 Bog Myrtle (*Myrica gale*)...15
2.3 Meadowsweet (*Filipendula ulmaria*) 16
2.4 Henbane plant (*Hyoscyamus niger*) 16
2.5 Henbane flower...16
2.6 The food vessel from North mains, Strathallan, Scotland, with Meadowsweet flowers 17

3.1 Katz and Voigt's theories of biocultural evolution 20
3.2 Map of the Near East and Levant to show sites discussed in the text 21
3.3 White ware vessels from pre-pottery sites in the Near East and Levant 22
3.4 Plaster statues from Ain Ghazal, Jordan, c6500 BC 23
3.5 The terracotta figurine of a goddess found in a grain bin at Çatal Hüyük, Anatolia......................................24
3.6 Wooden, stone and pottery vessels from Çatal Hüyük, Anatolia...25
3.7 Early Dynastic seal impressions depicting drinking scenes 26

4.1 Map to show the spread of agriculture from the Fertile Crescent into Europe and Britain.............................30
4.2 Map to show the areas inhabited by the Ertebølle and Linearbandkeramik cultures 30
4.3 Ground plans of House 59, Tell Ovchorovo, Bulgaria...31
4.4 Ceramic 'sieve' sherds from Linearbandkeramik sites 32
4.5 Linearbandkeramik bowl, cups, flint knives, arrowheads and polished stone adzes.......................................33
4.6 Characteristic Ertebølle pottery..34
4.7 Map to show sites of Swifterbant group settlements..35
4.8 Chronological chart to show European Neolithic cultures..36
4.9 Drinking vessels, baking plate and bowl from TRB or Funnel beaker culture 37
4.10 A so-called 'face pot' from southern Scandinavia 38
4.11 A variety of middle neolithic pottery found in passage graves, southern Scandinavia 38
4.12 Some of the types of Drouwen B TRB pottery from west Holland..39

5.1 Map to show the Orkney Islands with the location of sites discussed..41
5.2 Ground plan of the buildings at Knap of Howar, Papa Westray, Orkney 42
5.3 Large, deep pottery vessel from Knap of Howar 43
5.4 Wide, shallow bowl from Knap of Howar 43
5.5 The smaller building (House 2) at Knap of Howar showing the entrance 44
5.6 View of House 2 from beside the doorway, showing middle and rear sections 44
5.7 Areas constructed at the rear (left) of House 2 that may have been used for sleeping or resting.....................45
5.8 Storage places incorporated in the wall construction at the rear (right) of House 2..45
5.9 The passage leading from House 2 into House 1 to the right of the main entrance to House 2 46
5.10 The different sizes of doorways in House 1 at Knap of Howar 46
5.11 The low stone construction on the floor in House 1 47
5.12 The grain processing area at the rear of House 1 47
5.13 Ground plan of the settlement at Barnhouse, Orkney 49
5.14 Building 2, Barnhouse, showing drains in the northwest corner and by the entrance.....................................50

5.15   Ground plan of the settlement at Skara Brae, Orkney 51
5.16   Grooved ware pottery vessels from Skara Brae discovered during Clarke's excavations 52
5.17   Ground plan of Hut 8, Skara Brae 54
5.18   A good view of the kiln flue, Hut 8 55
5.19   Ground plan of Hut 7 showing furnishings as described by Childe 56
5.20   Inside Hut 7, looking at Pit P 57
5.21   The plinth by the door, inside Hut 7 57
5.22   Large pottery bucket from Rinyo, Rousay 58
5.23   Location map of Rinyo, Rousay 58
5.24   Ground plan of Rinyo, Rousay 59
5.25   Oven and hearth in Chamber C, Rinyo, Rousay 59
5.26   Oven base found under the floor of Chamber A, Rinyo, Rousay 59

6.1   Map of the British Isles, Ireland, Western Isles and Orkney showing locations of sites 64
6.2   Reconstruction drawings of two Grooved Ware vessels from Balfarg, Scotland, (P63 and P64) 66
6.3   Ground plan of Balbridie timber hall, Scotland 68
6.4   Ground plan of Lismore Fields, Derbyshire 69
6.5   Ground plans of selected timber buildings of Ireland 70
6.6   Ground plan and aerial photograph of Durrington Walls, Wiltshire 71

INTRODUCTION

"Our prehistoric fathers may have been savages, but they were clever and observant ones ... the art and practice of the brewer are founded on empirical observation ... the brewer learnt from long experience the conditions not the reasons for success"

John Tyndall, extracts from his speech on Fermentation
Glasgow Science Lectures Association
October 19th 1876

Grain in prehistoric diet

The preparation and consumption of food and drink are important aspects of prehistory that can provide a valuable insight into the daily lives of people in past societies. The introduction of the cultivation of grain in the Near East and the spread of the agricultural lifestyle across Europe and into the British Isles was a great change in the lives and habits of Mesolithic people. It is one of the most important changes to have occurred in prehistory. After millennia of subsistence activities based on hunting, gathering and fishing people began to cultivate and therefore to have control over a variety of crops, including wheat and barley. They also began to domesticate animals. Much has been written of this so-called "Neolithic Revolution", that is, the period of change from gathering, hunting and fishing to that of farming and herding. It was a change of lifestyle that occurred at different times in different parts of the world, but what was it that made people choose to cultivate wheat and barley, in particular?

Cereal grains are a major source of carbohydrate in the human diet, being useful for making porridge, bread and flour. They are also unique as a potential source of malt and malt sugars that can be fermented into beer or ale. With a minimum of simple equipment, such as containers, water and heat, it is possible to trick the barley into digesting itself into sugars. This aspect of grain processing has been overlooked in much of the archaeological literature relating to the transition from Mesolithic to Neolithic.

Brewing in the 21st century has become a global, multi-million pound technological business, with large breweries producing billions of gallons of beer annually. Many of these large breweries whose names are so familiar today such as Bass, Worthington, Younger and Guinness have only been in existence since the middle of the 18th Century. Prior to this quite recent industrialisation malt, beer and ale were manufactured either domestically or locally on a small scale.

The techniques of brewing small amounts of beer from malted grain have become largely neglected and the skill of domestic brewing is no longer a part of most peoples' daily experience. This thesis, based upon the biochemistry of malting and brewing and upon small-scale domestic brewing methods (Line 1980) proposes that Mesolithic cultures were interested in making particular products from the grain, that is, sweet malts and ale and that this was a major factor in the decision to selectively cultivate grain.

The 'bread or beer' debate

Robert Braidwood of the Oriental Institute of the University of Chicago (1953) first posed the question 'Did man once live by beer alone?' and this debate still continues today. Solomon Katz (1986, 1991) has coined the phrase 'biocultural evolution' and he argues for the importance of the transference of specialised food processing techniques to subsequent generations. Certain processing activities, such as brewing, become enshrined in ritual. Brian Hayden (1996) agrees with Katz and Voigt that grain was first domesticated to produce ale for consumption at feasts and at other special occasions. However, he notes the difficulty of finding direct archaeological evidence for early farming techniques and such grain processing activity as brewing (Hayden 1990).

Most recently Alexander Joffe (1998:297) has proposed "the production, exchange and consumption of alcoholic beverages form a significant element and regularity in the emergence of complex, hierarchically organised societies, along with the restructuring of labour and gender relations." Although these arguments are in the context of early Neolithic cultures in the Near East, the Levant and Egypt, they are equally as relevant to grain cultivation and processing across Europe and in the British Isles during the Neolithic.

Brewing in history and prehistory

Both the manufacture and the consumption of a wide range of alcoholic beverages are understood to have been important aspects of social, economic, religious and ritual life in Iron Age Europe (Dietler 1989), in Viking cultures and in early medieval Europe (Woolf & Eldridge 1994, Davidson 1998). Drinking horns and a huge bronze cauldron that contained the remnants of mead was found in a rich 'princely' grave at Hochdorf, Germany, dated to the 1st millennium BC (Biel et al 1985 Vol 1:147). A large quantity of carbonised malt, accidentally burnt as it was being kilned, was found at Eberdingen-Hochdorf (Stika 1996:81). Malt is the primary ingredient for beer or ale.

The earliest written references to ale being made in the British Isles can be found in the Vindolanda tablets, dated to the early 1st millennium AD. Roman soldiers recorded their purchases of barley ale made by the local tribes. Pliny refers to the Gallic tribes of Northern Europe making "intoxicating drinks from corn steeped in water...that are capable of being kept until they have attained a considerable age" (Pliny XIV Ch 29). There are also many references to the manufacture and consumption of ale and mead in the myths, legends and skaldic verse of the Viking Age in northern Europe.

Ale and mead were consumed on many occasions, for example at religious feasts and festivals, at funerals, in drinking competitions and before the men departed to sea in the spring (Gayre 1948:45, Davidson 1988:11,12). Women were usually responsible for the manufacture of alcoholic drinks in the societies cited above and there were close associations between the consumption of ale and the worship of deities (Dietler 1990:392, Joffe 1998:299, Davidson 1998:138). Ale is manufactured from malt, with herbs added for flavour and preservation. Mead is fermented honey and water with similar flavourings and preservatives as those used in the brewing of ale, such as Meadowsweet (*Filipendula ulmaria*). Honey was frequently added to the malt and so it is difficult to be clear as to the precise nature of the 'ale' and 'mead' referred to in ancient texts, myths and legends.

There is convincing evidence for the manufacture of

both ale and mead during the Bronze Age in Europe and in the British Isles. Organic residues within a beaker accompanying a female burial in a stone-lined cist at North Mains, Strathallan, Fife, were analysed and found to consist of cereal residues and Meadowsweet pollen. They were dated to c1540 BC (Barclay et al 1983). The excavators interpreted this as being the probable remains of a fermented cereal-based drink. At Ashgrove, Fife, Scotland, a beaker containing significant quantities of Lime Flower (*Tilia cordata*) and Meadowsweet pollen was discovered, again in a stone-lined cist accompanying a burial (Dickson 1978). The contents of the beaker were probably mead rather than ale. Vessels made of birch bark have been found at Egtved and at other Danish bog burial sites. Analysis of the contents indicates the "debris of wheat grains, leaves of bog myrtle (*Myrica gale*) and fruits of cranberry" (Dickson 1978:111). Bog myrtle was an additive used regularly as a preservative in the manufacture of ale prior to the introduction of hops in the late Middle Ages (Vencl 1994, Bennett 1996).

Neolithic Britain

In recent years organic residues that might indicate the manufacture of alcoholic drinks have been found on Neolithic pottery assemblages at ritual and domestic sites within the British Isles. Residues on sherds of Grimston-Lyles pottery and Grooved Ware from pits at Machrie Moor, Arran, were analysed and found to contain cereal pollen together with macro plant remains. These were interpreted as the probable remains of a mead-type drink (Haggerty 1991:91).

Cereal based residues were found on sherds of large Grooved Ware vessels that had been buried in pits situated close by a rectangular timber structure at a Neolithic and Bronze Age ceremonial site at Balfarg/Balbirnie, Tayside. Pollen from plants including Meadowsweet, Henbane, Deadly Nightshade, Cabbage and Mustards were noted in these residues, an interesting mixture of additives perhaps indicating some kind of fermented mead/ale type brew with special properties (Moffatt in Barclay et al 1993). At the Neolithic village at Barnhouse, Orkney, barley residues have been identified on some of the Grooved Ware vessels (Jones 2000). Scientific analysis, specifically Gas Chromatography and Mass Spectrometry, has indicated the presence of 'unidentified sugars' within the fabric of some of these vessels. These sugars might be maltose.

Thousands of charred cereal grains were found at the site of a large timber hall at Balbridie, Kincardine, dated to the early 4th millennium BC (Fairweather & Ralston 1993). Charred grain was also found at the site of a rectangular timber building at Lismore Fields, Buxton (Garton 1987). These finds and the cereal based residues described above are an indication of grain processing, perhaps for the manufacture of malts and ale, during the early Neolithic in the British Isles.

The possibility that grain processing activities during the early Neolithic of the British Isles included malting, mashing and fermentation should be considered and further investigated. Ian Hodder (1997:695) has argued for a destabilisation of 'taken-for-granted' assumptions in the interpretation of archaeological data and for the need to look at material culture assemblages as a complete whole. This multidisciplinary research and the subsequent interpretation of Neolithic grain processing techniques take this approach.

Brewing is "one of the oldest biotechnological processes of all" (Kretschmer 1996) requiring skill as well as specialised knowledge. Each stage of the process requires very specific and different conditions. In prehistory, the transformation of grain into malts and ale was very likely to have been an important social, symbolic and economic activity, as well as being a specialised and skilled craft that was passed on from one generation to the next. Malting, mashing and brewing have a great potential for apprenticeships, for the creation of social hierarchies and status and for the possession of secret or specialised knowledge. These grain processing activities may also have been extremely significant in terms of both ritual and social behaviour.

Andrew Sherratt has investigated and discussed the possibilities that drugs, such as cannabis and opium poppy seeds, were consumed in the Neolithic and Bronze Ages, perhaps as ritual or specialist activities (Sherratt 1991, 1995). Ale is also an intoxicant and a great deal of evidence exists for its manufacture and consumption during the Neolithic. There is also some tentative evidence for the ale to have been enhanced, at times, with psychoactive drugs such as Henbane and Deadly Nightshade although there is some contention and debate surrounding this issue (Long et al 1999). Whether or not alcoholic brews were enhanced with such additives is difficult to prove.

Malting and brewing in prehistory

In order to recognise the extant archaeological evidence for malting, mashing and fermentation it is helpful to understand the basics of the biochemistry as well as the methods and techniques of grain processing for malt sugars and ale. Chapter One examines the specific craft skills of the maltster and the brewer. Chapter Two examines some of the traditional and ancient use of herbal additives that preserve, flavour or strengthen the ale. The archaeological evidence for malting, mashing and brewing activity in the Levant, in the Near East and in Egypt is assessed in Chapter Three. Chapter Four assesses this evidence with respect to the European early Neolithic and Chapter Five examines the stone buildings of Neolithic Orkney in terms of grain storage and processing activities. Chapter Six investigates whether the Grooved Ware Culture of mainland Britain had a suitable material culture to make malt and ale from the barley grain that they grew.

Research for this thesis initially began with the Bronze Age of the British Isles. The original intention was to investigate the manufacturing techniques of Bronze Age brewers. However, barley has been cultivated in Britain since the early 4th millennium BC (Ashmore 1996). The focus of research soon turned to the Neolithic of the British Isles. In order to place British Neolithic grain cultivation and processing techniques into context it was necessary to look at the earliest development of cereal cultivation in the Near East, the Levant and Europe. The remit of this thesis has changed considerably as it has developed.

The Neolithic extends from the 9th/8th millennia BC in the Levant and Near East to the 4th/3rd millennia BC in the British Isles. This thesis covers a wide geographical area and an extensive timescale. It has not been possible to investigate every area in detail. Therefore selective sites have been chosen for analysis. This is an initial investigation into the possibilities for malting, mashing and brewing during the Neolithic.

CHAPTER ONE
The Conversion of Grain into Malt and Ale

Although people have been making fermented alcoholic drinks for millennia, it was only in the 19th Century that the scientific explanations for malting, mashing and fermentation were discovered. Yeast cells, responsible for the "enigma of fermentation" (McGee 1984:426), were first seen under the microscope in the mid 1830s. Two scientists, separately, observed the individual yeast cells splitting under the microscope: Theodor Schwann published his findings of the "new cell theory" in 1837 and Charles Caignard de la Tour had reported his observations a year earlier, in 1836.

In the 1850s Louis Pasteur (1822-1895) was asked by a Lille brewery to investigate the reasons for the spoiling of wine and beer, a serious problem that was costing the company many thousands of pounds. The science of Microbiology was a development of his groundbreaking investigations into the causes of fermentation and disease. Pasteur argued that yeast was a living organism that caused fermentation by chemical reaction (Conant 1952:22). His innovative research was continued by John Tyndall (1820-1893) who, when addressing the Glasgow Science Lectures Association in 1876, pointed out that "until the present year no thorough and scientific account was ever given of the agencies which come into play in the manufacture of beer, or the conditions necessary to its health, and of the maladies and vicissitudes to which it is subject" (Conant 1952:37).

Tyndall observed brewers' general practice of always excluding the air during the fermentation process. They used closed barrels or kept a lid or other suitable cover on the fermentation bucket. He knew that beer wort could ferment naturally in contact with 'common air', but that the product of this kind of fermentation was usually sour, extremely unpleasant and disagreeable to drink. His experiments, based on this knowledge from his observations, showed that only under anaerobic conditions, does the yeast plant "decompose the sugar of the solution in which it grows, produce heat, breathe forth carbonic gas and one of the liquid products of the decomposition is our familiar alcohol" (Tyndall in Conant 1952:41).

Rim diameter:	42cm / 17 inches
Height:	45cm / 18 inches
Volume:	45 litres / 10 gallons

26cm / 10 inches
28cm / 11 inches
11 litres / 2.5 gallons

Figure 1.1
Simple equipment can be used to ferment sweet barley wort. The wort is fermented in a brewing bucket with a lid and a smaller bucket is used for the transference of liquids. A lid keeps the airborne contaminants out of the brew and also creates the necessary anaerobic conditions for alcoholic fermentation (after Line 1980:164). Scale 1:10

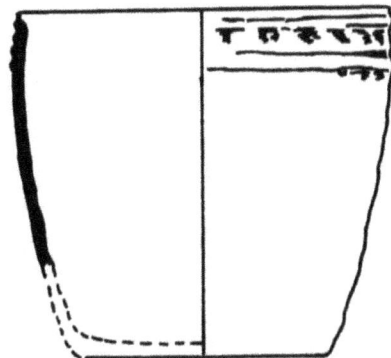

Rim diameter:	58cm / 23 inches
Height:	55cm / 22 inches
Volume:	115 litres / 25 gallons

36 cm / 14 inches
36 cm / 14 inches
32 litres / 7 gallons

Figure 1.2
Approximate dimensions and volumes of Grooved Ware vessels represented by sherds found at the Neolithic settlement at Skara Brae, Sandwick, Orkney (after Clarke 1976b). Scale 1:10

Anaerobic conditions are simple for the domestic brewer to achieve: beer or ale can be successfully fermented in a bucket with a lid. The carbon dioxide exhaled by the yeast plant is held, in concentration, as a layer above the fermenting wort, thus creating suitable conditions for alcoholic fermentation. A lid also prevents dust particles and bacteria from infecting the brew (figure 1.1).

Very large bucket-shaped Grooved Ware pottery vessels were found at Skara Brae. The sherds of one pot indicated that it might have measured 2 feet in diameter and 2 feet in depth, therefore having a volume of about 30 gallons (figure 1.2). This would have been a suitable vessel for the fermentation of a barley wort, provided that a lid was used to create the anaerobic conditions necessary for an alcoholic fermentation. Circular stone pot lids, suitable in size and shape, were found at Skara Brae, Orkney.

Pasteur's experiments into the causes of disease and fermentation provided a scientific explanation of the necessity for cleanliness and meticulous hygiene during the brewing process. Microscopes revealed the presence of bacteria and microbes that could be airborne in dust and that were capable of spoiling beer and other foodstuffs, such as milk and meat. By using dust-free chambers he demonstrated the existence of what he termed "the ferments of disease," microbes which sour the brew and infect food (Pasteur 1879). Cleanliness of all equipment is of paramount importance in brewing. Tyndall explained why an alcoholic fermentation could only take place when anaerobic conditions were present.

These ideas of hygiene and microbiology are now accepted but when presented to the scientific community of the late 19th century, they were contentious and controversial concepts. The scientific community of the time believed that bacteria and microorganisms were the result of 'spontaneous generation'. Tyndall made a convincing case for the then new and revolutionary hypothesis that fermentation was the result of the activity of micro-organisms rather than a spontaneous, mysterious and almost magical process, as had been believed previously.

The manufacture of an alcoholic drink from grain is a specialised craft that requires knowledge, experience and skill. Julian Thomas describes ritual as being "a form of human action which may involve a range of forms of material culture" (Thomas 1996:8). The conversion of the barley grain into malts and ale is such an activity. It requires a specific sequence of activities that lends itself to Thomas' definition of ritual. Brewing is an important aspect of history and prehistory and it is one that can be described accurately as a domestic ritual activity.

It is essential to understand the methods and techniques of the prehistoric maltster and brewer, for so long shrouded in mystery and secrecy, so that it is possible to recognise and interpret what is left in the archaeological record. For example, a quern or rubbing stone is usually interpreted as a tool only used for grinding grain into flour to make bread. It is equally efficient as a tool for crushing and cracking the malted grain for the better release of malt sugars when mashing.

A threshing or malting floor can be made of level clay or earth within a building, with frequent repair being an indication of it being used as a grain preparation surface. Large vessels are required to contain the wort as it ferments

and lids create the necessary anaerobic conditions.

1. The techniques and biochemistry of brewing

The craft of brewing is dictated by the biochemistry. The processing required to convert grain into ale is actually a sequence of three biochemical reactions, each requiring its own quite specific conditions. Malting renders the grain friable and much easier to grind. When mashed, the malted grain produces sweet 'cakes' or a sweet barley mash and a malt liquid that are rich in B-vitamins and an excellent food source. More sweet liquid (wort) can be washed (sparged) out of the barley mash, boiled up with herbs as flavourings or preservatives and fermented to produce an alcoholic liquor.

This section examines in detail how the sweet malts and ale are made from barley grain and what conditions and material equipment are needed to achieve this. Historical descriptions of processes are relevant, since the techniques are based on the same unchanging biochemistry since earliest times. There are some archaeological examples from later prehistory that illustrate the basic methods employed. Practical experiments in mashing and fermentation, using appropriate equipment, were carried out in order to properly assess the possibility that early Neolithic cultures were converting grain into malts and ale.

a) Malting: the controlled germination of grain.

Unprocessed barley is unsuitable for brewing because it contains only starch, which cannot be fermented. In order to make alcohol, this starch must be first converted into sugars which can then be fermented. This process begins with the germination of the barley grain (Line 1980:118). Grain is steeped in water for several days to trigger the growth mechanism. The water must be changed regularly in order to prevent spoilage of the grain and to provide oxygen. A simple and effective method of doing this is to put the grain in a sack and leave it in a stream for two or three days (Flett, H. pers comm). If a stream is not available and the grain is to be steeped in a vessel, then the water must be changed daily to maintain the freshness of the grain. After this, the grain is spread out in a layer 4-6 inches deep on the malting floor and regularly raked, turned and tended. It must be kept in a warm, dark environment so that the grain produces a rootlet and a shoot, known together as the acrospire, and as the grain grows the starch is converted into malt flour.

These changes are triggered by enzymes, molecules that living cells use to transform other molecules (McGee 1984:426). The enzymatic process is a complex one. There are many different enzymes that are activated by germination and which carry out specific functions within the grain, for example, cytase, proteolase and amylase. (Line 1980:119). Maltsters today control the malting process with precision, allowing the seedling to grow until the acrospire is about two-thirds the length of the grain. Current techniques involve spreading "a six inch layer of barley on the floor of a large room, where it can be easily raked to keep it aerated and moist but not wet" (McGee 1984:471).

Ancient malting techniques were exactly the same, with the barley grain being spread out on a malting floor, preferably within a dimly lit building that would provide shelter from the elements and protection from birds and animals. The grain was watered and raked at regular intervals to prevent the grain from drying out or moulds from developing on grain that became too wet. Close observation and care of the grain's growth are essential at this stage. In the 'Hymn to Ninkasi', an ancient Sumerian hymn of praise to the Goddess of Brewing that is dated to the mid 2nd millennium BC, mention is made of watering the malt and of the 'noble dogs' guarding the precious malt and keeping away 'even the potentates' (Katz & Maytag 1991:29).

According to descriptions of the 9th century AD, Ireland, malt takes between 12 and 15 days to produce. The process involves 24 hours of steeping, 36 hours draining, four and a half days under cover and three days lying exposed until "it is heaped up in piles, then raked or combed into ridges before being finally dried in a kiln" (Comey 1996:21). Variations occur according to local climate and season.

A malting floor is a level surface that may be made of beaten earth, clay, plaster, stone or wood. Shelter from the elements and protection from domestic and wild animals are essential since the malt is an attractive food source. Over years of use, the floor would require repair, being re-plastered if made of plaster, or the addition of newly packed layers of earth or clay, if that was its original base.

b) Kilning the malt

If grain is allowed to grow unchecked then the starch will be used up, so the grain must be dried out or kilned. This terminates germination and dries the malt so that it can be stored until needed (Line 1980:120). Dry malt keeps well. Malted grain is easier to grind than unmalted. For the mashing and fermentation experiments, described below, pale crushed malt purchased from a brewing suppliers was used. In later experiments, it was possible to use Bere Malt that had been grown on Orkney and malted by Harry Flett at the Corrigall Farm Museum (figure 1.3). Bere barley is now only grown in a few fields in Scotland and Orkney, although it is slowly re-gaining its popularity. The Bere barley grain was malted in a barn on an earth malting floor that is over 200 years old and, according to Harry Flett, in need of some repair and attention. The surface was broken and it needed to be smoother to allow for raking and turning the malt.

The grain barn at the Corrigall Farm Museum, Orkney

The grain barn at the Corrigall Farm Museum is a stone building with a flagstone and turf roof. It was built around 200 years ago and is still used today. It is a building with many functions including the storage and repair of tools and farm equipment as well as the threshing, winnowing, malting, kilning and storage of grain.

The grain could be winnowed between the two opposing doorways, making use of the through draft created by the wind (Harry Flett pers comm). The grain is threshed, winnowed and then malted on the earth floor before being spread out to dry in the kilning area that is constructed at the rear of the barn (figure 1.4). The kiln has a drying area above and to the right of it, not directly above the flames and heat, accessible by two stone steps built in the wall and just visible in the photograph. The malted grain is spread out on hay or straw that is laid over some chicken wire. It dries gently in the warm airflow from the kiln.

Figure 1.3
Inside the grain barn at the Corrigall Farm Museum, Harray, Orkney. Bere barley used in experimental mashing work was malted on this floor by Harry Flett, a curator of the museum. The 200 year old malting floor is made of beaten earth. The grain drying facilities can be seen at the far end of the barn with the kiln on the left and the grain drying area accessed by stone steps, to the right.

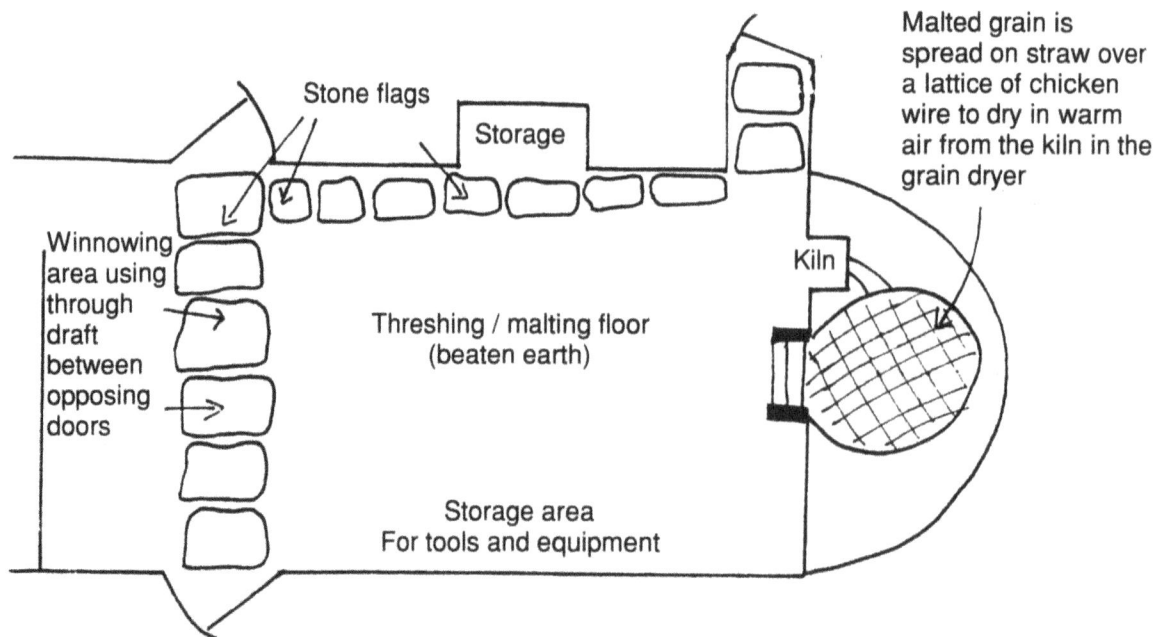

Figure 1.4
Rough ground plan of the 200 year old grain barn at the Corrigal Farm Museum, Harray, Orkney (not drawn to scale).

Iron Age malting

Evidence for Iron Age malting has been found at Eberdingen-Hochdorf in southwest Germany, where remains of a brewery dated to between 600 and 400 BC have been excavated (Stika 1996:81). Here, malted grain was dried on a lattice of sticks laid over pits with fires being lit beneath, the grain drying gently in the hot air (figures 1.5,1.6). The unusual preservation of "weakly but evenly germinated charred barley grains, enough to indicate malting," were discovered in 6 long U-shaped ditches that were 5-6m long, 0.6m wide and 1m deep. Wooden boards supported the sides of the trenches. Fires lit at the ends of the trenches allowed the malt to be gently dried after germination. It appears that the fire became out of control and burnt the entire contents of the trench. According to the excavator the germination was deliberate and controlled, the grain had been threshed and cleaned prior to germination and all the evidence points to malting and drying of the malt, a necessary precursor to ale production.

Zosimus, writing in the late 3rd or early 4th century AD, explained that in hot dry climates malt can be dried in the sun, provided that it is protected from being eaten by birds (Lucas 1962:14), hence the guardianship of the malted grain by dogs in the Hymn to Ninkasi as discussed earlier. Traditional malt drying methods in Scotland, perhaps dating back into prehistory, include rolling hot stones in the germinated barley or drying malt on hot flat stones by a fire. (S.E.A.). Dried malted grain is friable and contains starch, malt flour, maltose and dormant starch-converting enzymes.

c) Mashing: the conversion of starches into malt sugars.

Crushing the malt prior to mashing is crucial. This allows the enzymes to intermingle with the starchy endosperm and there is a better conversion of starch to malt sugars during the mash. This crushing naturally produces malt flour.

The crushed malted grain is mashed so that all the starches are converted into sugars, a process known as saccharification. Modern and medieval techniques of mashing involve 'striking' the malt with hot, almost boiling water to achieve a final temperature of between 65 and 67 degrees Centigrade. Temperature is critical at this point. If the water is too hot the starch converting enzymes will be killed. Too cool a temperature and the enzymes will not re-activate optimally.

From Iron Age times onwards mash tuns constructed of wooden staves have been used. Once the technology of making stave-built containers had been mastered it was possible to make very large buckets and barrels. Prior to this, wooden vessels had to be carved out of solid wood, thus limiting the potential size (Earwood 1993). An interesting method of putting hot stones into wooden mash tuns continues today, as in the manufacture of *Steinbier* or Stonebeer (Kretschmer 1998). The hot stones heat up the barley mash slowly and saccharification takes between 3 and 5 hours, provided that the temperature never becomes too hot to kill off the enzymes.

Slavomil Vencl (1994) describes archaic practices of mashing barley malt that have been preserved in Europe until recent times: "... the technique of brewing in wooden vessels with the aid of fire-heated stones is a relict, not an abberration (Maurizio 1927:26)" (Vencl 1994). In Carinthia, this century, stones were heated in a fire of cherry wood (to add flavour to the liquid) and water with malt was heated in this way until it caramelised (Hopf and Wiegelmann 1976:532)." (Vencl 1994:310). Such an activity would leave few archaeological traces, requiring only a wooden vessel, a fire, stones, water and malted barley. Malting and then mashing barley and other grains to extract the sugars, using either wooden or pottery bowls, was probably one of the earliest grain processing techniques practiced in prehistory.

During mashing, the enzymes act as a catalyst to bring about a chemical reaction that reduces the remaining grain starch to fermentable sugars. Diastase converts the starch into a mixture of sugars grouped as dextrins and maltose, with the dextrins being less fermentable. Diastase is actually a grouping of two separate enzymes: alpha and beta amylase. Enzymes respond to the temperature and acidity of the mash differently.

4

Figure 1.5
The excavated remains of the U-shaped ditch at Eberdingen-Hochdorf, N. Germany, which was used for drying the malt. It was 5-6m long, 0.6m wide and 1m deep with wooden boards to support the walls. Within the pit was a structure of dried mud bricks and a wooden frame, which may have been covered with textiles or reeds. The malt was spread out on this and dried in warm air generated by a fire at one end of the pit. An accident probably caused the destruction of this kilning pit which contained burned mud bricks, charcoal and charred grain. (Stika 1996:66)

Figure 1.6
Carbonised malt from the pit at Eberdingen-Hochdorf. The remains of the burned wooden lattice support are visible among the malted grains. (Stika 1996:67)

Alpha amylase converts starch into dextrins and is tolerant of higher temperatures than beta amylase, which converts the dextrins into maltose (Line 1980:125/6).

Mashing is complex in terms of enzyme activity but it is very simple to do. Egyptian and Sumerian brewers made *bappir*, a kind of flat barley 'cake' or 'bread' (Samuel 1995). The slow rising temperature activates the enzymes, which convert all the remaining grain starch into sugars within the first few hours. *Bappir* store extremely well, if kept dry.

Mashing experiments

In the first experiments I used Fawcett's pale crushed malted barley, purchased from a home brewers' suppliers. In later experiments I used Bere malt that was grown and floor malted by Harry Flett, curator of the Corrigall Farm Museum, Harray, Orkney. The crushed and malted barley was mixed with copious amounts of water in a waterproofed pottery bowl and heated gently in the warm ashes of a fire (figure 1.9). The bowls that were used had been proofed with beeswax, which seals the porous fabric of the pot and makes it watertight.

Great care was needed to ensure that the enzymes were not killed. This involved constant observation of the mash and the regular monitoring of its temperature. The temperature of the mash was checked using a brewing thermometer and also checking by touch. The correct 'feel' for the temperature for best saccharification was soon learnt, and there was no need for a thermometer. The optimum temperature for the enzymes to convert starch into sugars is between 65 and 67 degrees Centigrade (Line 1980).

Pale crushed malt and water were combined to make a fairly stiff mixture for the barley 'cakes'. These were then placed on a hot flat stone to cook. They needed to be kept moist, as they dried out quickly and the enzymes require moisture to work. It was also necessary to keep the fire burning fairly fiercely, to provide a constant supply of hot ashes to keep the mash temperature correct and the flat stone hot. At times, the mash became too hot and had to be taken

out of the fire. It soon became apparent that the hearth was a little too small for mashing in this way (figures 1.10, 1.11). Nevertheless, the mash sweetened as the gentle heating progressed. After about an hour the saccharification could be seen, smelt and tasted in the bowl mash, the mixture turning brown and some of the malt becoming caramelised and burnt onto the side of the pottery bowl.

The liquid was tasted regularly. It began to taste sweet after about an hour and it became increasingly darker, sweeter and stickier as the mashing progressed. The barley 'cakes' were much slower to saccharify, as it was difficult to maintain a high temperature of the stone without over-heating the barley mash in the bowl.

In subsequent experiments, a larger hearth made entirely with flat stones was constructed, based on those found at Neolithic settlements on Orkney (figure 1.11). This was a far more functional hearth than the earlier one used. It was much easier to run a fire and to cook at the same time in a hearth of this size. The cooking area was far enough away from the area where the fire was allowed to blaze, providing hot ashes but, critically, not overheating the mash. The stones by the blazing fire became very hot, radiating heat when the fire had been running for some hours. They proved to be excellent for making barley 'cakes'. It was not difficult to imagine the potential heating capacity by radiation of these stones should a fire be kept going in the hearth over several weeks, months or years, as was the case at Skara Brae in Neolithic times.

Given that about five pounds of barley mash were needed to make the two gallons of wort for later fermenting, it was not practical to mash these large amounts over an open fire. As a practical expedient, I used the domestic oven. The pale crushed malt was mixed with water and placed in cup-cake trays in a very, very low oven for several hours. This guaranteed a slow rising temperature for a good conversion by the enzymes of grain starches into sugars. There was a sweet and powerful aroma when the correct temperature for saccharification had been reached, the same aroma as noted when mashing over the open fire and when the barley malt is 'struck' with hot water using modern wet mashing techniques.

After oven mashing, the result was a dark brown, sometimes caramelised and crisp mix of barley husks and malt sugars (figure 1.12). Because the temperature was more precisely controlled, the conversion was excellent, with strong caramel flavours and some burnt grains. The mixture tasted very sweet and was extremely sticky to handle.

The mash was stored in every large, deep vessel available in the house and it was kept covered to avoid contamination whilst the later batches were being made. It took several days of oven mashing to produce the five pounds of mash needed for two gallons of wort. When the mash was left to stand, it was noticed that a quantity of pure malt liquid, very dark brown in colour and tasting very sweet, had filtered through the barley husks to the bottom of the container (figure 1.13). This malt liquid was sweet and delicious. It is a useful and nutritious end product of malted and mashed grain and it contains easily digestible B-vitamins.

Photographs have been included of the mashing experiments that I did in 1998. One of the particular aims of including these colour photographs is to demonstrate the dramatic change from a pale, white, starchy mixture of malted barley and water to a dark brown, sweet-tasting and caramelised mashed barley mixture with malt liquid. This transformation is easily achieved over the gentle warmth of hot ashes or, alternatively, in a low oven. As a novice at this I struggled to maintain the fire, to provide sufficient hot ashes and to monitor the temperature of the barley mash, all at the same time.

In the past, grain processing would probably not have been a solitary task. In my experiments it became clear that any attempt to make large amounts of barley mash, sufficient for several gallons of ale to be produced, would have involved a group of people working together. In the experiments, approximately five pounds of malted barley was mashed to produce about two gallons of sweet wort for fermentation. The process took several days. The next section explains how the sweet liquid wort is extracted from the mash, leaving 'spent grain', which is used as animal fodder.

d) Obtaining a sweet wort for fermentation

It is easy to rinse out more of the sweet and sticky malt liquid from the barley mash, a practice known in the brewing trade as sparging, with the product being the wort (figure 1.14). Hot water is used since "sugar, when in solution, will flow much more easily and readily when it is hot. Sweet wort is no exception, and the whole theory of sparging is based on this principle...the husks act as a filter." (Line, 1980:141*ff*). Usually, the amount of water used to sparge is approximately equal to the desired gallonage of beer.

In my experiments, hot water was trickled through the mash using a basket-sieve over a large container. This is the method depicted in Egyptian brewing scenes of the 3rd millennium BC (figure 1.8). The method worked well, but was a very slow and extremely sticky and messy business. Modern methods and equipment were used, specifically a grain bag, two plastic buckets and a plastic syphon tube (figure 1.7). Two gallons of sweet wort was thus obtained.

The grain and chaff eventually settled out to the

bottom of the bucket, and further careful pouring eliminated much of this chaff from the wort until a clear liquid was obtained for fermentation. Syphons may be practical solution, plastic ones being used in my experiments. In a prehistoric context, reeds might function well. Holes and spigots at the base of vessels would also work very well but there would probably be a lot of spillage of the wort.

McGee (1984:474) notes that "fully 85% of the carbohydrate in malt was starch while in the liquid wort 70% is now in the form of various sugars: small amounts of glucose, fructose, some 3-sugar remnants of starch and about 40% maltose, or two-glucose molecules". The wort is very sticky indeed, and the photographs give a slight indication of the sticky mess that always accompanies the handling or transferring of the wort from one vessel to another. Vessels and equipment require constant washing. Access to a nearby water supply and to drains is crucial for processing grain in this way.

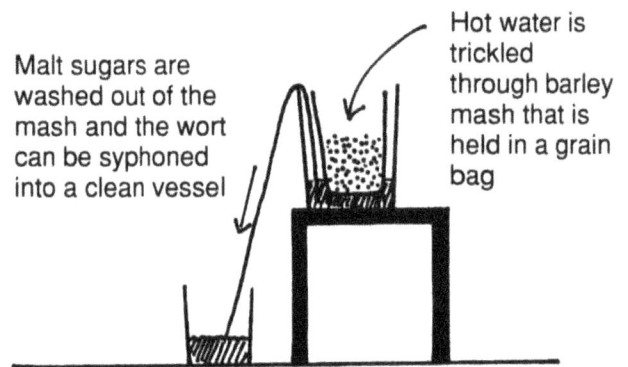

Figure 1.7
Sparging the wort using a grain bag, a bucket, a table and a syphon. The wort is collected and then allowed to settle before the next stage.

Figure 1.8
Egyptian maltsters, mashers and brewers c1500 BC from a painting on the wall of a Theban tomb. The processes of raking the malt, making bappir, sparging and fermentation can be seen (Singer 1980: figure 180).

Mashing experiments using crushed malted barley

Figure 1.9
Pale crushed malted barley was mixed with abundant water in a bowl that had been waterproofed with beeswax. The bowl was then placed in the hot ashes of an open fire. Initially, at this stage, the liquid was opaque and white, showing it to be starchy at this stage prior to the mashing. The malted barley was very pale brown in colour.

Figure 1.10
As the malt and water gently heats up the enzymes within the grain re-activate and continue the conversion of starch into sugars. Saccharification is indicated by the darkening of the barley mash and at this stage there is a powerful aroma that is strong and sweet. The mash changes in colour and it gradually becomes dark brown and tastes sweet after 15-30 minutes. Some of the malt becomes carbonised. The mixture that had been placed on the hot stone needed to be constantly wet for the saccharification to occur.

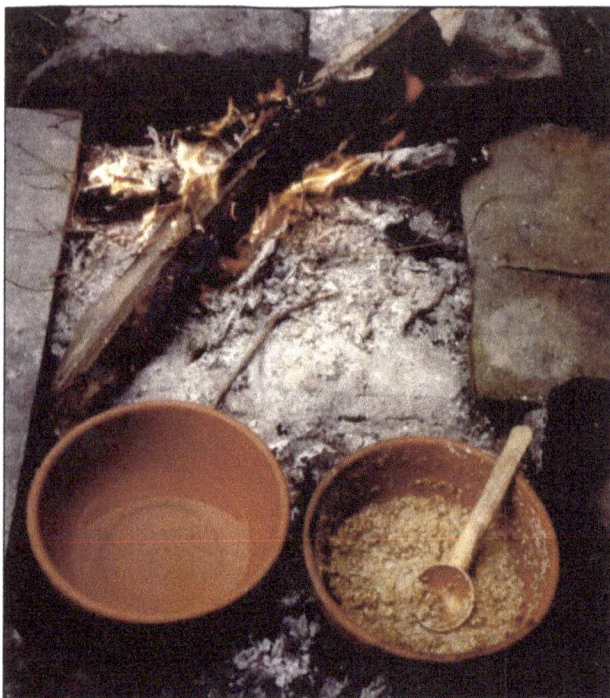

Figure 1.11
The larger rectangular hearth based on Orcadian Neolithic hearths excavated at Skara Brae and Rinyo.

Figure 1.12
Three bowls, one containing crushed malted barley and two with sweet barley mash, showing the contrast in colour.

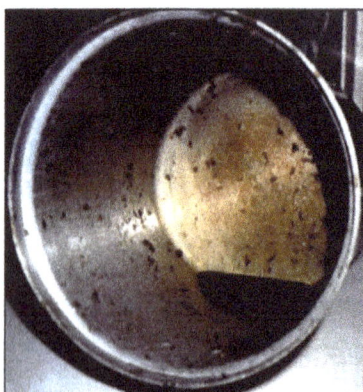

Figure 1.13
Pure, sweet malt liquid has filtered through the mash and husks to collect at the bottom of the vessel used for temporary storage of the barley mash. This malt liquid is highly nutritious, being rich in B-Vitamins. It can be used in several ways. If added to milk it makes a delicious malted milk drink, just like Horlicks. It can be added to other foods or dishes as a sweetener. The mash can also be sparged using hot water to extract more malt sugars (the wort) that can be fermented into ale.

Figure 1.14
More wort can be collected by sparging the mash with hot water. This involves running hot water slowly through the mash and collecting the run off in a bucket. This illustration shows the wort when first extracted and after one hour, having been left to settle. The cloudy mixture settles out, leaving a clear sweet malt liquid (wort) at the top and chaff and other bits at the bottom of the vessel. After settling the wort can be transferred into a clean vessel, then boiled with the desired additives for flavour and preservation and then, finally, it is fermented. This is an extremely sticky and messy business. All equipment must be washed to prevent contamination of subsequent brews. Cleanliness of equipment is crucial.

e) Boiling the wort with flavourings and preservatives.

Boiling the wort kills the enzymes that are still active and which, if left unchecked, will spoil the ale (Line 1980:157). It also sterilises the wort and it precipitates out proteins that affect the flavour and storage of the beer or ale. Herbs that are added to the boil can either add flavour to the brew and or they can act as a preservative. Hops, cultivated in Europe since the eighth or ninth centuries AD, perform both these functions but in prehistoric times other herbs were used, such as Bog Myrtle, Ground Ivy and Meadowsweet (Mabey 1996:64,317). Aspects of the use, the purposes and properties of a selection of herbal additives to ale other than hops is examined in Chapter Two.

Meadowsweet pollen was found in significant quantities in the residues of the Strathallan Beaker and those found in the Ashgrove Beaker (Barclay et al 1983, Dickson 1978) and so it was decided to use this particular flower as an additive in these brewing experiments (figures 1.16,1.17). Meadowsweet requires boggy and damp conditions to thrive and it is frequently found growing along ditches, by lakes and streams and in wet meadowland. Meadowsweet flowerheads were gathered on Orkney, where it grows profusely, and were dried for use in these experiments. Some was gathered at Loch Harray, close by the Neolithic settlement at Barnhouse and the Stones of Stenness. The meadowsweet was a very well established plant there, with thick, tough, woody stems that required cutting with a knife. Care was taken not to take too much meadowsweet from any one area and so it was gathered, in small amounts, at a number of different locations on mainland Orkney. It propagates by the root system; the different flowerheads being part of the same plant. Picking the flowers does not prevent the plant from thriving (figure 2.3).

In order to properly assess its flavouring qualities, two separate brews were made from a batch of malt liquid obtained by oven mashing and sparging. In one brew of about a gallon, a fifth of an ounce of dried meadowsweet flowerheads was added during the boil. A second brew, also of about a gallon was boiled but no herbs at all were added. Both were fermented in glass demijohns with fermentation locks and, when fermentation was complete, were stored in plastic bottles as a practical expedient, although fermentation in pottery vessels with a suitable lid is a possibility and will be done in future experiments.

It is essential to seal the porous fabric of the pot with either beeswax or fats to prevent the contents leaching out through the walls of the vessel during fermentation. Experiments in this area are ongoing.

After one week, the brew without any dried meadowsweet flowers added to the boil had become sour, very nasty and undrinkable. The brew that had been boiled with added meadowsweet flowers remained fresh enough to drink for several months afterwards.

f) Fermentation: the conversion of sugars into alcohol.

There are several ways that yeast can be introduced to the wort. Wild yeasts exist in the air, as noted by Pasteur and Tyndall. In Belgium today, lambic beers are fermented using wild yeasts alone, a relict perhaps of ancient practices. Alternatively, a yeast culture can be kept alive and used when needed, as modern breweries do today. Suitable yeast cultures for brewing may have been cultivated and kept 'alive' in prehistory. Yeast can survive, in a dried state, on the surface of pottery vessels. For example, in Egypt, such residues dating to the 2nd millennium BC have been found and examined using a scanning electron microscope (Samuel 1995,1996).

It is most unlikely that such residues have survived in Northern Europe and the British Isles as they have in the dry climate of Egypt. The addition of sweet wort to a pot with yeast residues dried on to its internal surface would have the effect of activating the yeast and fermentation would then begin, as if by magic. Yeast requires a temperature of between 55-70 degrees Fahrenheit, but it metabolises best at around 60 degrees. The yeast action splits the sugar into alcohol and carbon dioxide gas. (Line 1980:164). Anaerobic conditions are necessary for alcoholic fermentation, the alcohol being a 'waste product' of the yeast, which is forced to survive by using the sugars as a source of energy:

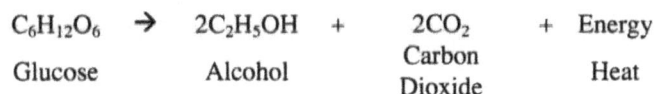

$$C_6H_{12}O_6 \rightarrow 2C_2H_5OH + 2CO_2 + Energy$$

$C_6H_{12}O_6$	$2C_2H_5OH$	$2CO_2$	Energy
Glucose	Alcohol	Carbon Dioxide	Heat

Energy is a by-product of the biochemical process. The fermenting wort appears to boil and creates its own warmth in the absence of a fire. Although the biochemistry of yeast activity is a new discovery, its effects are clearly visible. At first, the bubbles of carbon dioxide rise up slowly, but as the yeast action increases over time, the mixture appears to boil. According to David Line "after 12-24 hours the yeast crop can look quite frightening. Long meringue-like tentacles of yeast reach out from the surface of the brew, some reaching 6-8" in length. The contents of the dustbin look more like a Quatermass experiment than beer in the making" (Line 1980:163). This subsides after 8 hours and over the next 3-4 days the fermentation quietens down and is less spectacular.

The size of the fermentation vessel used depends upon the amount of wort to be fermented. The experiments in mashing and sparging produced small volumes of wort, therefore one gallon demijohns had to be used, together with airlocks. Larger quantities of wort could be easily fermented in buckets with lids (figure 1.1). The meadowsweet ale obtained from these experiments was a clear, dark reddish-brown in colour, with a flavour of meadowsweet. Very few adverse comments have been received from the tasters.

Meadowsweet Ale has been made for the Manchester University Archaeology Society to sample in December 1997 and some was made for the Neolithic Studies Group Meeting in March 1998. Some is still keeping well, in plastic bottles in a cool cellar. To leave the ale in contact with the yeasty sediment would soon taint it, so, when fermentation is finished, the ale must be either syphoned off or carefully poured into a clean vessel for consumption or storage. This process is referred to as 'racking off' the ale. Plastic syphon tubes were used to syphon the ale made as part of this research into clean plastic bottles. As noted earlier, reeds would probably work efficiently as syphon tubes, but experiments in this have not yet been tried.

The residues that accumulate at the bottom of fermenting vessels differ from barley mash residues, known as 'spent grain' or from residues resulting from sparging the wort, in that they also contain pollen and macro plant debris from the herbs added as flavourings or preservatives during the boil. Therefore, organic residues that are described as 'barley mash with added herbs or flavourings' probably represent the residues of a fermented drink, either manufactured or stored in that vessel.

Once the sugars have been washed out, little would remain of the 'spent' barley mash. It makes an excellent feed for both cattle and pigs and it is not a waste product that would remain in the archaeological record.

2. Necessary materials and equipment

Barley and wheat are versatile crops that can be processed in several ways to make flour, bread, porridge, malts and ale. Most of the literature assumes grain to be a source of carbohydrate in the diet and that it was grown as a staple food for its flour, porridge and breadmaking potential. The fact that grain can easily and with a minimum of equipment also be processed to extract sweet malts has been overlooked. Grain was probably processed in a variety of different ways in prehistory.

This research concentrates specifically upon the potential in the early Neolithic for the manufacture of sweet malts and ale from barley and wheat. The basic requirements for malt and ale production are grain, water, large vessels, fuel and yeast, if the malt is to be fermented. All types of grains can be used such as wheat, barley, oats and rye (Vencl 1994:307). Slavomil Vencl's analysis of the basic requirements for brewing can be expanded upon in the light of evidence from this chapter. A malting floor, made simply from earth, clay, wood or stone is needed. Shelter and protection from the elements and from birds and beasts, in the form of a roofed and enclosed structure, is another very important factor. Hearths, ovens or kilns are necessary for drying out and for mashing the malted grain (figure 1.15).

Access to a nearby water supply is essential for washing the used sticky pottery vessels and also for sparging, in which the sugars are washed out of the mashed barley using heated water. Cleanliness is crucial to brewing and a strong source for ritual behaviour. Suitable pottery is required, both large and small vessels for storage, for fermentation and for drinking from.

Without doubt one of the most crucial aspects of brewing is the skill and knowledge of both the maltster and the brewer in processing the grain correctly in order to make a successful brew.

Grain	Most grains are suitable, for example, wheat, barley or rye.
A malting floor	Smooth, level floor of beaten earth, clay, plaster, wood or stone.
A barn	The malting floor should be situated in a dark, well-ventilated building to maintain an even temperature and to protect the germinating grain (the malt) from the elements, birds and beasts.
Water	Mashing and fermentation require copious amounts of water. All equipment needs to be kept scrupulously clean to avoid contamination of the product.
Drains	Useful for transporting waste water out of the building.
Kiln, oven or hearth	To dry the malt and to mash the malted grain.
Containers	To mash the malted grain, to ferment the mash and to store the products. They may be pottery or wood. If wooden, hot stones are necessary to heat the mash.
Cover for vessel	To provide anaerobic conditions for the ferment and to exclude airborne contamination.
Sieves	For the separation of the wort from the mash.
Yeast	Needed to start the ferment. Can be cultivated in a pot but will survive, dried, on the internal surfaces of the fermentation vessel or on a hazel wand, used to stir the brew.
Knowledge and skill, practice, experience	With the above equipment and ingredients, it is possible to make ale from grain. The skill exists in knowing the right way to do it.

Figure 1.15
Table showing the basic equipment required for processing grain into malt, malt sugars and ale. Knowledge, skill, practice and experience are important factors, being as important as the materials and equipment.

	Sample 1 Scraping from pot	Sample 2 Black greasy deposit
TREES		
Pine (*Pinus*)	0.2	
Birch (*Betula*)	0.2	0.2
Oak (*Quercus*)	0.4	0.2
Hazel (*Corylus*)	3.2	1.0
Alder (*Alnus*)	1.1	0.2
HERBS		
Heather (*Ericaceae*)	0.2	
Ribwort (*Plantago Lanceolata*)	0.7	0.3
Sedges (*Cyperacaea*)		1.2
Grasses (*Gramineae*)	8.1	4.0
Hawthorn (*Rosacaea*)	0.2	
Flax (*Linum Usitatissimum*)		0.2
Cruciferae	0.2	
Buttercup (*Ranunculaceae*)	0.4	0.9
Compositae	2.7	3.7
Meadowsweet (*Filipendula*)	66.6	74.1
cf Filipendula	8.1	8.0
Cereals (*Cerealea*)	1.4	0.3
FERNS		
Polypodium	0.4	0.5
Monoleet psilate-type	2.9	0.7
MOSSES		
Sphagnum	0.4	
Indeterminate	2.7	4.9

Figure 1.16
Pollen analysis of organic 'black greasy' deposit and scraping from inside the Bronze Age food vessel (SF17) found at North Mains, Strathallan, Scotland. There are relatively high values of meadowsweet pollen and this indicates its deliberate rather than accidental addition to the contents of the pot (Barclay et al 1983).

Small Leaved Lime (*Tilia cordata*)	53.7	Scabious (*Scabiosa*)	0.2
Meadowsweet (*Filipendula ulmaria*)	15.1	Grasses (*Gramineae*)	0.2
Heather (*Calluna*)	7.7	Willow (*Salix*)	0.2
Ribwort Plantain (*Plantago Lanceolata*)	7.1	Beech (*Fagus*)	0.2
Mint (*Labiatae*)	5.1	Oak (*Quercus*)	0.2
Alder (*Alnus*)	2.8	Sphagnum	+
Holly (*Ilex*)	2.2	Cereals	+
Daisy (*Compositae*)	1.2	Pine (*Pinus*)	+
Hazel (*Corylus*)	1.1	Birch (*Betula*)	+
Buttercup (*Ranunculaceae*)	0.8	Honeysuckle (*Lonicera*)	+
Polypody Fern (*Polypodium*)	0.8	Red Shank (*Polygonum*)	+
Ferns (*Filicales*)	0.5	Persicaria	

+ = Found On Scanning Slide

Figure 1.17
Pollen analysis of Ashgrove Beaker from Scotland, with very high values of lime. Heather and meadowsweet pollen were identified in smaller quantities (Dickson 1978).

The contents of this pot may have been mead (fermented honey) rather than ale.

CHAPTER TWO
Additives used to Preserve, Flavour and Enhance Ale

1. Additives used in Medieval and Viking times

Prior to the introduction of Hops (*Humulus lupulus*) in late medieval times a wide variety of additives have been used by brewers to preserve, flavour, strengthen or clarify ale. Bog Myrtle (*Myrica gale*), Ground Ivy (*Glechmona hederacea*), Mugwort (*Artemesia vulgaris*) and also Meadowsweet (*Filipendula ulmaria*) were among the most popular and the most commonly used herbs by medieval brewers and alewives (McGee 1984:466; Davidson 1998:154, Genders 1971:174). Hops were introduced in Europe around the 9th century AD and in the British Isles in the 14th century.

Hops were not favourably received at first by the populace of the British Isles. Ale and beer were considered to be separate drinks, ale being the traditional drink made with local herbs and beer being made with imported hops. The manufacturers of both drinks maintained their separate Guilds for some time until eventually, by the 16th century, beer became the predominant drink. Hop growers from the continent settled in Kent during the 15th and 16th centuries, founding the now thriving Hop business there (McGee 1984).

In medieval and ancient times herbs would have been added to the wort during the boil as leaves, as fresh or dried flowerheads or as whole plants if their function was as flavouring or as preservative agents. If the clarification of a cloudy ale was required, then they would be added to the fermented ale. The cultivation and gathering of herbs and the knowledge of herbal properties, either medicinal or culinary, were of great importance in prehistoric and early historic times. The herb garden was an essential part of house and home. It provided plants useful for cooking purposes as well as plants to use as remedies for illnesses and the healing of wounds (Davidson 1998:154 *ff*). It is difficult to make a clear distinction between the domestic, culinary, medicinal, ritual and functional use of herbs in the past.

Herbal remedies can be powerful and potent, even fatal, if taken incorrectly or in excess. They are certainly not risk-free cures and the knowledge of their correct usage was and still is a complex skill. Herbal remedies are becoming increasingly popular in the late 20th century. The Department of Complementary Medicines at Exeter University, led by Professor Edzard Ernst, is currently undertaking research into their clinical efficacy.

In Viking and Medieval times brewing was one of the domestic responsibilities of women (Davidson 1998:138*ff*) and it would be fair to assume that this was probably also the case in prehistory. It was customary for women to brew ale at home until the middle of the 17th century AD, when the process became industrialised. Malting and brewing began to be practised on a large scale when the very first breweries were established. Brewing ale domestically was a labour intensive business involving much hard work. It provided women with an income and ale was often sold on by alewives and brewsters of the Middle Ages (Bennett 1986, 1996).

a) Ground Ivy

In medieval times and earlier the leaves of Ground Ivy, also known as Ale-Hoof, were commonly used in ale making (figure 2.1). The natural habitat of this herb is in woodland and beneath hedgerows, as it grows best in partial shade. It was a commonly cultivated herb in domestic gardens and in those of wayside inns (Genders 1971:174).

Nicholas Culpepper (1616-1654), the astrologer-physician, noted that that Ground Ivy could clarify a cloudy ale 'in a night, that it will be fitter to be drank the next morning; or if any drink be thick with removing or any other accident, it will do the like in a few hours.' The herb is 'sharp and bitter to the taste' and it has several healing properties, for example, an infusion 'easeth all griping pains, windy and choleric humours in the stomach, spleen or belly' and a decoction of it 'helpeth wounds...ulcers...scabs, weals and other breakings out in any part of the body' (Culpepper's Complete Herbal).

b) Bog Myrtle

Bog Myrtle is a small shrub that grows prolifically in bogs, wet heath lands, fens and moors (figure 2.2). It is currently found mainly in Scotland, North Wales and northwest England (Rose 1991:234) but in prehistoric and early historic times it could probably have been found in suitable environments throughout the whole of British Isles.

The shrub "emits a resinous, balsamic fragrance, especially when in flower.... and the whole plant is still used in brewing and cooking.... being added to home made beer by mixing the leafy branches with the hot liquid in the early stages of beer making" (Mabey 1996:70). The leaves of the plant and the stem are covered in aromatic glands, that "release a refreshing resinous scent when handled" (Genders 1971:129).

c) Mugwort

Mugwort is another aromatic herb that was frequently used as flavouring or medicinal additive in domestic brewing in medieval times and earlier (Genders 1971:200; Davidson 1998:154). It was considered to be a 'women's herb' being useful in childbirth. According to Culpepper it had the effect of hastening the delivery and also helping to expel the afterbirth. It was used extensively in home medicines, being valued in many parts of the world and it has been suggested (Armstrong 1943, quoted in Davidson 1996:154) that it could have been one of the earliest deliberately cultivated herbs.

d) Meadowsweet

Meadowsweet is a tall aromatic plant that thrives in damp and boggy places, by rivers, in marshland, in meadows and in ditches. It was regularly used in historical times to flavour and preserve mead, hence its name, a derivative of 'mead-sweet' (Mabey 1996:181). As described in Chapter One, the dried flowerheads of this herb were used to flavour and successfully preserve the ale that was made as part of this research. Strictly speaking, 'ale' should refer to fermented barley wort and 'mead' to fermented honey and water. However, alcoholic beverages "were often brewed with both honey and malt, and the literature refers to these variously as mead, ale or honey ale" (Ratsch 1994:280).

Meadowsweet was also used as a strewing plant, having a 'pretty, sharp scent' that 'far excels all other strewing herbs, to deck up houses, to strew in chambers, halls and banqueting houses' (Genders 1971:87). Culpepper notes that an infusion of the flowers is 'good for all fevers...and a good wound herb, whether taken inwardly or externally applied' (Culpepper's Modern Herbal, Foulsham edition: 230). This is due to the fact that meadowsweet contains salicylic acid, which reduces pain and fever. It is interesting to note that herbs useful to the brewer as flavourings and preservatives also have medicinal properties. In ancient times, women were responsible for the cultivation, gathering and preparation of herbs that were used for many aspects of home medicine, healing and nursing of the sick and during pregnancy and childbirth. Women were responsible for the general management of house and home, including plant cultivation and food preparation (Davidson 1998:138,154).

One of the purposes of the previous chapter was to demonstrate the potential for a 'domestic ritual' aspect to the craft of brewing, that is, the necessary and essential sequence of specific actions and processes which must be adhered to. With the customary addition of a variety of herbs to the ale, herbs that may have a medicinal or psychoactive effect, there is added a further dimension to this ritual activity. Wise women and healers in antiquity and prehistory had the specialised herbal knowledge and the skills of herb usage to help and to heal people within their community.

e) Henbane

Plants with medicinal, aromatic and psychoactive properties, such as Henbane *(Hyosycamus niger)* and Deadly Nightshade *(Atropa belladonna)* were at certain times added to ale or mead (Ratsch 1994:285). Henbane (figures 2.4, 2.5). and Deadly Nightshade are highly poisonous plants of the Solanacae family. They have long been used for medicinal and ritual hallucinogenic purposes. Henbane contains tropane alkaloids, specifically hyoscyamine, hyoscine and atropine. Modern medicine still uses derivatives of Henbane for travel sickness and atropine is a useful drug in the treatment of Parkinson's disease.

Hyoscyamine is "a powerful hallucinogen .... which gives the sensation of flying through the air.... among many other effects" (Devereux 1997:98). Henbane is a plant well known for its ability, when ingested in limited amounts, to induce delusions of flying. Demented mental states, confusion and hallucinations are also effects produced by the plant when ingested. Henbane had developed strong associations with magic and witchcraft practices by the early Middle Ages in the British Isles and in Europe (Sherratt 1996:14). It is possible to achieve these hallucinogenic effects by inhaling the smoke given off when the seeds are burnt, by rubbing ointment containing extract of Henbane on the skin or by drinking ale with Henbane added to the brew (Lehane 1977:184). Sometimes herbs are added during the fermentation rather than the boil because their alkaloids are alcohol soluble rather than water soluble.

Henbane has a powerful, pungent aroma described as 'a very heavy, ill, offensive smell' by Culpepper, who was writing in the mid 17th century (Culpepper:184). More recently it is described as the 'unpleasant, sickly, fetid smell of dead rats' (Genders 1971:151). The whole plant, including

the root, which looks like a parsnip, is highly poisonous. There is no antidote. Culpepper advises that: 'The herb must never be taken internally, it is altogether an outward medicine.'

The leaves could be applied externally to relieve headache or local swelling. The 'oil and juice of the herb or seed' could be used as eardrops - 'good for deafness, noise and worms in the ears.' A decoction of the herb or seed 'kills lice in man or beast.' According to Culpepper 'the fume of the dried herb, stalks and seeds, burned, quickly heals swellings, chilblains or kibes in the hands and feet, by holding them in the fumes thereof.' Gerard (1595) notes that the fumes of Henbane seeds, when inhaled, were an effective cure for toothache. Both Gerard and Culpepper wrote of common medical practices of the late 16th and early 17th century. Pliny associated Henbane with death and wrote that it was used in funeral meals and was scattered over tombs (Lehane 1977:146). It must be likely that there was knowledge of these and many other herbal remedies in prehistoric times.

2. Archaeological evidence: herbal additives in prehistoric Europe

There is interesting archaeological evidence that indicates the probable use of Henbane as an additive to ale during the first millennium BC. A total of 15 Henbane seeds were found in association with charred and malted barley grains at the Late Hallstatt/Early La Tene site at Eberdingen-Hochdorf, which is situated 15km northwest of Stuttgart and has been dated to c500 BC. The site was interpreted by the excavator as being a brewery as well as being a settlement with special status. The discoveries at this site are most unusual and "the extraordinary nature of beer and mead production at Eberdingen-Hochdorf should be stressed" (Stika 1996:88).

It is clear from finds at the rich grave, close by the brewery at Hochdorf, that ale and/or mead were being manufactured and consumed in very large quantities in Iron Age Europe, although not always with Henbane as an additive. A huge bronze cauldron capable of containing 500 litres of liquid was found in a rich burial (Biel et al 1985). An 8-10 mm layer of sediment at the bottom of the cauldron had "a pollen content so great that the honey, the source of the pollen, would have sufficed to produce an alcoholic drink - mead" (ibid: 147).

The effects of Henbane ale are vividly described by the Muslim traveller and writer Ibn Fadlan, an Arab expert in religious law (Ratsch 1994:279). In 921 AD he had been sent by the Caliph of Baghdad to instruct the people of the northern Bulgars in the Islamic faith. In his writings, he describes the funeral ceremony of a Viking chieftain of the Rus, who were Scandinavian merchant adventurers who regularly visited Russia to trade in furs and slaves. The chief had died at Bulgar, an important trading centre on the banks of the Volga and funeral rites took place there for him, according to Viking traditions. Specially brewed Henbane ale was drunk day and night by the participants until they became demented and some even drank themselves to death (Ratsch 1994:285). This ale was called 'nabid' and was an intrinsic part of Viking funeral rites, together with human and animal sacrifice (Davidson 1998:164-166).

Figure 2.1
The picture (left) shows Ground Ivy, a plant commonly known as 'ale-hoof' (Glechmona hederacea). It is found growing in woods, in hedgerows and on damp, rough ground. It was used in ancient times for herbal infusions and as a bittering agent for ale (Mabey 1996:317).

Figure 2.2
The picture below shows Bog Myrtle (Myrica gale) growing in damp heathland on Hartland Moor, Dorset. It is a shrub of the wet acid heathland and moors of Scotland, North Wales and Northwest England. The whole plant is used in brewing, being added to the boil as a flavouring and as a preservative (Mabey 1996:70).

Figure 2.3
Meadowsweet (Filipendula ulmaria) in flower. It thrives in damp riverside meadows and most other damp and boggy places, such as ditches and the shores of lakes. It was much used in historical times to flavour and to preserve mead, hence its name, which is derived from 'mead-sweet'. The plant flowers in July and August. The flowers have a rich and powerful scent, making it excellent as a 'strewing plant'. It propagates by its root system, which spreads along the ditches. It is a tough and hardy plant that is found in abundance in Scotland and Orkney (Mabey 1996).

Figure 2.4
Henbane (Hyoscyamus niger) has grey-green leaves densely covered with sticky hairs. The flowers are pale yellow with purple veins. It grows in sandy places, by the sea and sometimes on disturbed chalk lands. This plant (left) was found growing at Lulworth, Dorset. The hallucinatory and poisonous properties of the plant were well known in medieval times (Mabey 1996:301)

Figure 2.5
The Henbane flower (Mabey 1996).

Figure 2.6
The food vessel from North Mains, Strathallan, with meadowsweet flowers. Traces of meadowsweet pollen were found in this vessel together with cereal pollen and cereal-based residues. The vessel was discovered in a stone-lined Bronze Age cist that contained a female burial (Barclay et al 1983, Clarke, Cowie and Foxon 1985:202).

At times, the herbal additives used in such 'special brews' must have been as important or perhaps even more important than the ale itself, given their "hallucinogenic, aphrodisiac or psychoactive properties" (Ratsch 1994: 279).

Several crushed Henbane seeds were recently found in the same context as an organic residue described as "a lump of carbonised porridge" (Sherratt 1996:14) adhering to a sherd from a very large Grooved Ware vessel. The pottery had been deliberately deposited in a pit by a Neolithic timber mortuary enclosure at the ritual site at Balfarg/Balbirnie, Fife, Scotland. It could either indicate a medicinal use of the seed (Barclay et al 1993) or it could indicate the ritual practice of adding Henbane to ale. The cereal residues are described as being "ritually charged material" and are interpreted as 'porridge' (ibid 1993:109). They are much more likely to be the remains of sediments left by the fermentation of a barley wort into ale. The Henbane seeds would have been added to the brew during the boiling stage in order to make it more potent with added hallucinatory effects. This possible interpretation is discussed in more detail in Chapter Six, where the practicalities and possibilities of the manufacture of malts and ale during the British early Neolithic are summarised and assessed.

Henbane ale, if consumed in small amounts, would have very obvious intoxicating and hallucinogenic effects. If consumed in large quantities it would induce dementia, hallucinations and ultimately death. Such potentially dangerous additives as Henbane were probably not regularly added to ale or taken on a regular basis, but were a part of special ritual events only. Sherratt (1991,1995,1996) argues for the ritual or medicinal consumption of a number of narcotic substances during the Bronze Age and the Neolithic, such as the opium poppy *(Papaver somniferum)*, Henbane *(Hyoscyamus niger)* and the hemp plant *(Cannabis sativa)*. He cites the discoveries of opium poppy seeds at the Neolithic lake villages, Neuchatal, Switzerland and also at Linearbandkeramik sites in the Rhineland. He argues that the use of psychoactive substances in prehistory is "fundamental to the nature of sociality and an active element in the construction of religious experience, gender categories and the rituals of social life" (Sherratt 1995:34). From the evidence at Balfarg/Balbirnie it appears that such activities and rituals occurred during the British Neolithic.

It is generally accepted that ritual and social activities that included "feasting, from domestic celebration to communal occasions, requiring the large scale slaughter or sacrifice of animals and the brewing of drinks" (Sherratt 1991:51) were a significant aspect of Bronze Age life, both in Europe and in the British Isles (Burgess & Shennan 1976). This is supported by the analysis of cereal-based residues from the Strathallan Beaker, radiocarbon dated to c1540+/-60 BC (figure 2.6). The food vessel or beaker accompanied a female burial in a stone-lined cist at North Mains, Strathallan, Fife. A mixture of cereal residues and Meadowsweet pollen within the beaker was interpreted by the excavators as being the probable remnants of fermented cereal-based drink, namely ale (Barclay et al, 1983).

A beaker that was found in a stone-lined and clay-sealed cist grave dated to the middle Bronze Age at Ashgrove, Fife, Scotland, (Dickson 1978) contained plant debris and pollen from immature meadowsweet *(Filipendula ulmaria)* flowers and from flowers of the small leaved lime tree *(Tilia cordata)*, indicating that the beaker had contained mead. Vessels made of birch bark containing the residues of a mead or ale type drink have been found at Egtved, Denmark, and at other Danish Bronze Age bog burial sites (Thomsen 1929, ref to in Dickson 1978). The contents of one vessel, described as a 'birch bark bucket' by Professor Gram in the late 1920s, contained "debris of wheat grains, leaves of bog myrtle and fruits of cranberry" (Dickson 1978:111).

There is, then, significant and convincing evidence for the consumption of ale and mead during the Bronze Age. These alcoholic drinks were probably made by the women who were the principal gatherers, cultivators and food processors within the community. Arguments have been made for Bronze Age "male drinking cults" (Burgess & Shennan 1976) but there is no real archaeological evidence for this. The Strathallan Beaker, for example, was found in the cist grave of a female. Beakers have been found accompanying both males and females, indicating the probable consumption of alcoholic drinks by both genders. Women may have been the maltsters, the brewers and the ale manufacturers. It seems very unlikely that they did not consume the product that they so carefully and skillfully made.

Ale made prior to the introduction of hops was a very different drink compared with the gaseous, carbonated beer that we are now accustomed to, even though it is fundamentally the same product and is made according to the same unchanging biochemical laws. The various herbal additives would make the ale taste quite unlike hopped beer. Ale was a flat, not a fizzy, carbonated drink. Carbonisation was an invention of the mid 20th century and it was introduced into Britain in the 1940s by American servicemen during the Second World War.

The use of Hops as a preservative and flavouring was first promoted by the Christian monks of the Medieval period and, although popular in Europe from the 8th or 9th centuries AD, hops were not used throughout the British Isles until the 15th or 16th century (Ratsch 1994). The use of herbs in ale, particularly Henbane, had close associations with paganism and with witchcraft practices and would understandably have been a practice that the Church greatly wanted to discourage.

The material culture of the Neolithic is very similar to that of the Bronze Age. At both periods of prehistory there were buildings suitable for grain storage and processing, and suitable large pottery vessels for food and drink storage and for mashing and fermentation purposes. The early Neolithic cultures grew barley and wheat for the products that could be made from them, not only the flour and bread but also the sweet malts and the ale. In Egypt and Syria during the 4th and 3rd millennia BC, ale and other alcoholic beverages were an important part of ritual, domestic, commercial and economic life. The next chapter examines the first cereal and grain cultivation and processing activity in the Levant and in the Near East. It assesses the possibility that these early Neolithic groups were processing the wild barley and wheat into sweet malts and ale and cultivating grain deliberately as a source of sweetness.

CHAPTER THREE
Barley in the Levant, Ancient Near East and Egypt

1. The 'bread or beer' debate

The question whether barley and other cereal grains were cultivated for the manufacture of bread or of beer has been an ongoing debate since 1953. Robert Braidwood of the Oriental Institute of Chicago asked whether "the discovery that a mash of fermented grain yielded a palatable and nutritious beverage acted as a greater stimulant toward the experimental selection and breeding of the cereals than the discovery of flour and breadmaking?" (Braidwood 1953:515). He was attracted to the idea that the manufacture of beer or ale was the main motivation behind early grain domestication and he invited the opinions of other archaeologists and anthropologists in a now famous symposium that was published in American Anthropologist.

Jonathon Sauer agreed with Braidwood's idea that "thirst rather than hunger may have been the stimulus behind the origin of small grain agriculture" (ibid:516). Hans Helbaek strongly disagreed, pointing out that carbonised rather than malted grain had been found at Jarmo interpreting this as the baking of bread. Paul Mangelsdorf noted that cereals were the only carbohydrate furnishing food source available to early Neolithic people. He felt it to be highly unlikely that "the foundation of Western Civilisation was laid by an ill-fed people living in a state of intoxication" (ibid:520). Leo Oppenheim believed that the quest for food, storage and the different processing techniques of that food were instrumental in the discovery of both bread and ale (ibid: 521). This seems to be a reasonable compromise of the several arguments and it is the most likely explanation. The probability is that bread, malt, and subsequently ale were made in the early neolithic Near East.

Zohary and Hopf point out "when charred slowly and mildly, wood, seed, nuts, and sometimes even fleshy fruits or whole ears of cereals can still retain most of their morphological and anatomical features" (ibid:4). Such features are preserved "with astonishing clarity". At high temperatures carbonisation causes certain "characteristic deformations" of grain, such as shrinkage in the length of the kernel, 'puffing' of the kernel around the circumference and/or cracking of the grain (ibid: 4). Therefore it should be possible to examine the internal structure of ancient carbonised grain using scanning electron microscopy to try to ascertain the level of heat to which it was subjected. Malting requires very gentle warmth. Archaeological evidence of well-preserved charred grain might be indicative of malt burnt accidentally during kilning.

The debate generated by Braidwood's symposium still continues 50 years later. Solomon Katz and Mary Voigt (1986) have pointed out the dietary disadvantages of unprocessed wheat and barley. Unprocessed grains are not nutritious. Cereal grains are made up mainly of carbohydrates with only 13-20% protein and very low amounts of fats, B-vitamins, minerals and lysine. Lysine is an essential amino acid that enables the human body to process the other amino acids of the grain into proteins. As yeast grows in ale and bread it "produces a rich source of lysine, significantly

improves the B-vitamin content of the mixture.... thereby permitting the absorption of more essential minerals such as calcium" (Katz and Voigt, 1986:30). In short, this means that processed grain contains digestible B-vitamins. It is more nutritious than unprocessed grain.

In prehistory, people were not aware of the complex biochemistry that is necessary for the improved nutritional value of cereals. But they did observe that malting (partial germination and subsequent drying) made the grain taste sweeter and make it more palatable. Malting has the advantage of making the grain friable and therefore much easier to crush or grind up. As explained in Chapter One crushed malted grain, when gently heated in a bowl or oven with copious amounts of water, will always produce a sweet barley mash with malt liquid because of the enzyme activity. Unmalted grain can only produce a starchy porridge or gruel.

Having mashed the malt, it is but a few steps from the first, perhaps accidental, fermentation of the sweet malt liquid by wild airborne yeasts to the successful control and management of the whole malting, mashing and fermentation process. Observation, skill and practice are all that is required to learn how to manage these processes. Knowledge of the complex biochemistry is not necessary. Ale has been made successfully from the grain for millennia, long before the complexities of the biochemistry were understood and explained by scientists.

Katz and Voigt's theories of 'biocultural evolution' and the development of 'cuisine' centre on the ways that people process foods in these highly complex ways to "transform marginally nutritious and outright toxic substances into high quality foods" (Katz & Voigt 1986:25) (figure 3.1). Food processing techniques are learnt by one generation and passed on to, perhaps even improved upon by the next generation. Katz makes the point that "this information transfer requires stability. In all probability, myths, stories and legends, some of which are woven into ritual practices, all play a critical role in the process of stabilising the content of traditions that are passed from one generation to the next, as well as the social context through which they flow. Once stability has been achieved, the trial and error process that must originally have led to the evolution of the specific traditions is no longer necessary" (Katz & Voigt, 1986:25). The most crucial and important rituals and traditions relate to the acquisition, processing techniques and consumption of food (ibid: 25).

The conversion and transformation of barley grain into malts and ale is a set of activities that very easily lends itself to the acquisition of complex ritual practices and to the transfer of these practices through the generations. The manufacture of malt and ale from the grain comprises a set of traditions, rituals, skills and knowledge that could accurately be described as a domestic ritual activity.

2. Hunting and gathering groups in the Near East and Levant: 9th / 8th millennia BC

Modern species of six-rowed barley (*Hordeum vulgare* subsp *polystichum*) have been developed and selectively bred by agriculturalists over millennia from its wild ancestor, the wild two-rowed barley (*Hordeum vulgare* subsp. *spontaneum*). Wild barley is a natural plant of the area known

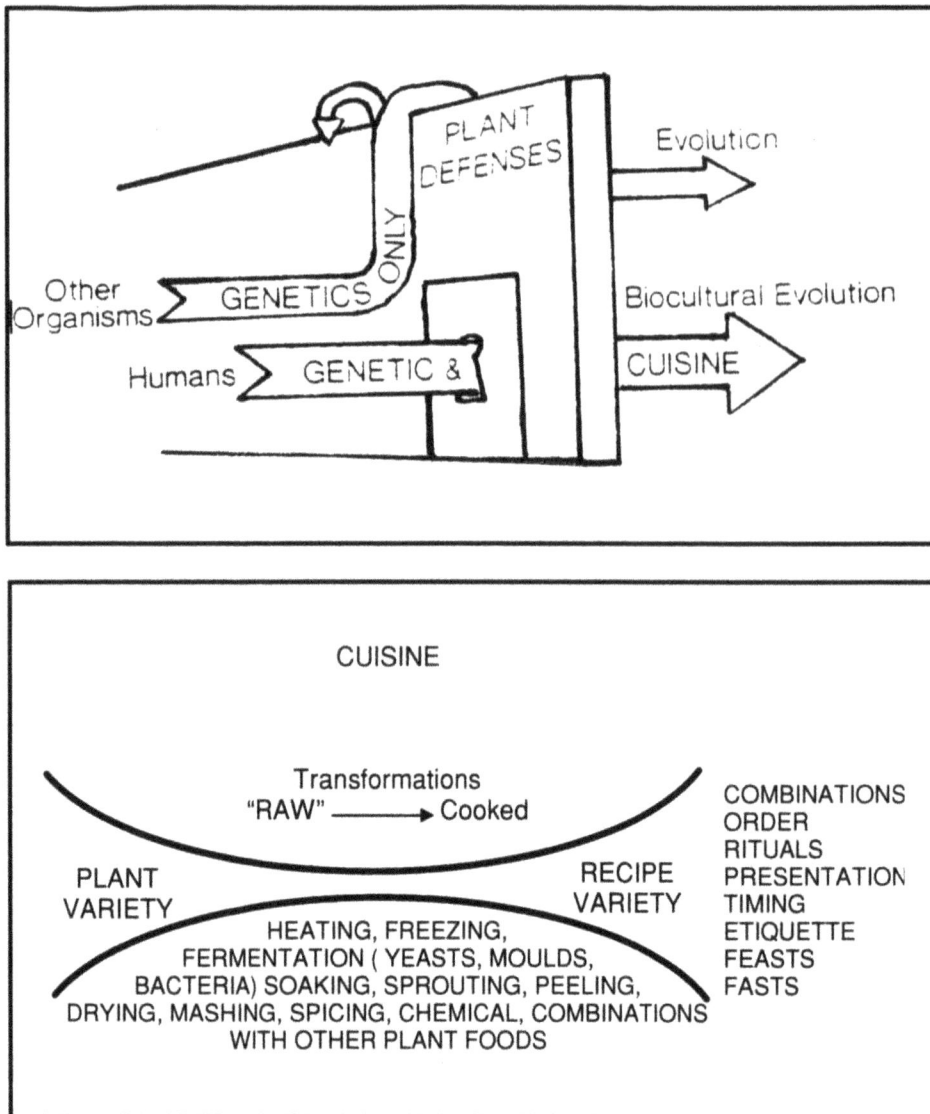

Figure 3.1
Katz and Voigt's theories of biocultural evolution and cuisine are best explained by the diagrams reproduced here. Plant defenses are eventually overcome by evolution but this is a slow process. Humans can 'unlock' these defenses by a combination of cultural and biological adaptations, by creating a set of traditions and ritual food preparation techniques that is a biocultural phenomenon termed 'cuisine' by Katz and Voigt. Cuisine is defined as the transformation of raw or modified plants or plant products into cooked or culturally modified foods (Katz and Voigt 1986:24).

as the 'Fertile Crescent'. This includes the modern countries of Israel, Jordan, southern Turkey, Iraq, Kurdistan and southwest Iran. Barley grows prolifically in the Jordon Rift Valley and the Tigris/Euphrates basin and sporadically in the Aegean region, around the Mediterranean coast, central Asia and Afghanistan (Zohary and Hopf 1988:58).

A number of sites were occupied during the 9th and 8th millennia BC in the Levant, the area now known as Syria, Jordon and Israel, for example Tell Abu Hureyra and Mureybit, both in north Syria and Tell Aswad, near modern Damascus (ibid:60). Sickle blades have been found in the earliest levels at these sites. Some were even made from baked clay. Wild grain seeds were stored in pits. The native species of wheat and barley were gathered by hunter-gathering groups, known as the Natufian culture (Katz & Voigt 1986:32). Jack Harlan, visiting an area by the Karacadag Mountain in the province of Diyarbakir, southeast Turkey, during the 1960s has described "vast seas of

primitive wild wheat" (Harlan 1967:197). He conducted harvesting experiments, collecting the wild grain by hand and also using a flint sickle. With the flint sickle he gathered 2.45 kg of grain within an hour and he estimated that, in a three-week period, enough grain to last a family for a year could be gathered easily and quickly. Unprocessed grain is not a good nutritional food source and so it had to be processed in some way.

The Natufian culture is usually seen as "a transitional phase between two cultural sequences, these being the Palaeolithic hunter gatherers on the one hand and the Neolithic agricultural societies on the other" (Belfer-Cohen 1995:9). These people lived a nomadic life. They gathered wild foods, hunted birds and animals, fished and moved around the area seasonally in order to exploit a wide variety of natural food resources, amongst which were wild barley (*Hordeum spontaneumi*) and einkorn of two varieties (*tritium boeoticum* var *aegilipoides* and var *thaudor*) which grew abundantly in the area (Thorpe 1996:6). The material remains of the early Natufian culture precede the earliest food-producing groups of the Neolithic Near East and Levant (Byrd & Moynahan 1995). The culture can be divided into two or three developmental stages, based upon technological and stratigraphic criteria (Belfer-Cohen 1995). It is not clear when the Natufian people first learned how to convert grain into malt sugars. They were either gathering wild grains or beginning to cultivate them by c8500 BC (Maisels 1990). It is likely that they were making several products and that malting was an early discovery. Allowing the grain to grow a little before grinding or crushing with mortars makes it a much easier task. This is because enzymes activated during the germination process break down the husk of the grain and make the starchy endosperm friable. This process naturally produces malt flour.

Tell Abu Hureyra, northern Syria

Tell Abu Hureyra is situated in northern Syria, by the Euphrates, about 40 kilometres from Mureybit. The skeletal remains of 162 individuals were found (Moore 1979). These human remains span 3000 years of the site's occupation and have been analysed by Theya Molleson at the Natural History Museum (Molleson 1994). Activities that were regularly carried out by the inhabitants have left their marks on the bones. Molleson discovered that bones of the upper spine were deformed, indicating the carrying of heavy loads on the head. Skeletal evidence indicates that grain was regularly pounded in pestles and mortars or crushed using querns.

According to Theya Molleson, the introduction of pottery technology c6000 BC coincided with an increase in the numbers of dental caries and decay in the teeth of the Abu Hureyra people. She attributes this to "a greater emphasis on cooked cereals made into bread and porridge...sticky foods that adhere to the teeth and provide the medium for the growth of bacteria that cause caries" (ibid: 64). Similarly high levels of periodontal disease and dental caries have apparently been noted at the Natufian site of Nahal Oren (Smith, P. 1989, 1991, noted in Thorpe 1996:6).

Today sugars are a well-known cause of caries and dental decay. Perhaps these people had discovered the processes of malting and mashing the grain and the manufacture of malt sugars. As shown in the experiments described and illustrated earlier malted barley, when heated gently with water, always produces not a bland, starchy gruel or porridge, but an extremely sweet and sticky barley mash, together with liquid malt sugars.

Malt is nutritious and it is an excellent source of easily digestible B-vitamins. The addition of B-Vitamins to the diet would have improved the health and probably the life expectancy of all inhabitants, both adults and children alike.

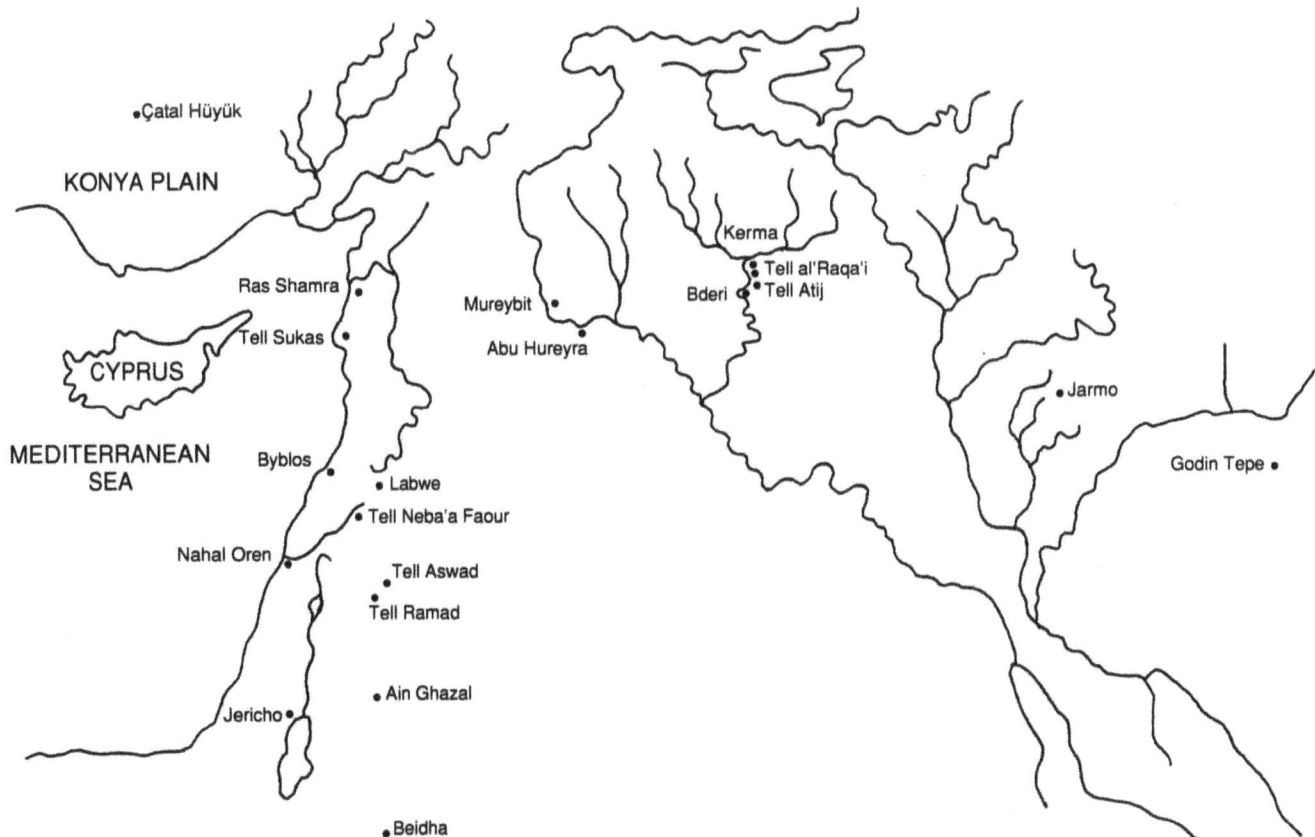

Figure 3.2
Rough map of the Levant and Near East, showing Neolithic and Bronze Age sites that are discussed in the text.

3. Early grain cultivation and processing in the 7th/6th millennia BC

Cultivated varieties of both barley and emmer wheat have been found during excavations at aceramic Neolithic sites of the late 8th and early 7th millennia BC, for example at Tell Aswad, Jarmo, Jericho and Ali Kosh (Zohary & Hopf 1988:60). The seeds differ from their wild counterparts, suggesting the beginning of deliberate and selective cultivation and domestication. The presence of grinding stones at these settlements indicates that grain processing was an activity. It is not clear what end product was being manufactured. Malt is a possible and likely candidate, since querns and grinding stones are as useful for crushing malt as they would be for grinding unmalted grain into flour. It has been suggested that "the event that 'primed the pump' and led people to invest energy in the collection and propagation of wild wheat and barley was the discovery of new food processing techniques - the sprouting and fermentation of these grains" (Katz and Voigt 1986:27). Malting and mashing require only a malting floor, a hearth, oven or kiln, hot stones and containers. Fermentation of the wort only requires a large container with a suitable cover to keep airborne contaminants out of the ferment. Yeast and access to a water supply are also necessary.

A few small ceramic vessels were found in early levels at Mureybit dating to c8000 BC, namely two cups with flat bottoms, a cylindrical vase and a small oval bowl, together with clay-baked female figurines and artefacts described as "batons" (Maisels 1993:85). Maisels considers these to be genuine ceramics and they are the earliest so far found in the Near East. There have been a few finds of White Ware, or Vaiselle Blanche. This is a precursor to clay-fired pottery and is described as "a composite of lime...and salty grey ashes. Vessels, often of a large size, were built up in coils round a basket...and when dried and fired this White Ware turns into a hard white material resembling limestone" (Mellart 1975:62). Although White Ware is not pottery in the true sense of the word, it would probably have functioned in a similar way to pottery vessels (figure 3.3).

White Ware is dated to between the second half of the 7th millennium BC and the first half of the 6th millennium BC. Remains of White Ware have been found at Ain Ghazal in eastern Amman (Rollefson 1983,1984, 1986), at Tell Sukas in Syria (Riis & Thrane 1974:26), at Tell Ramad II, close by Mount Hermon, Syria, (de Contenson 1971:278) and at Byblos, on the Lebanese coast (Mellaart 1975) (figure 3.2).

As well as the manufactured White Ware vessels described above, pre-pottery cultures also made vessels and containers from stone. At Beidha, a pre pottery Neolithic village dated from c7000 BC and situated high in the mountains south of the Dead Sea, the people used stone bowls, troughs and mortars, with baskets coated with bitumen and lime plaster also serving as containers (Maisels 1993:87). Such vessels would be suitable for the containment of liquid products from grain processing, for example the malt sugars and the ale. Strains of wild barley were cultivated and access to a reliable water supply would not have been a problem since the settlement is situated close by the spring of Dibadiba, in Wadi Gharab (ibid:79).

At Ali Kosh, dated from c7200 BC, the evidence points to a community that cultivated grain and also gathered the wild strains of wheat and barley. Wild game was hunted. The floors of the houses were made of stamped mud, often surfaced with a layer of clean clay and then sometimes overtopped with reeds. There is evidence of regular repair and maintenance of floor surfaces.

The inhabitants built domed mud brick ovens, they made stone mortars and used "stone bowls made of limestone or marble...most likely for the gruel." (Maisels 1993:101). The product is more likely to have been not 'gruel' but liquid malt sugars. The mud and clay floors would have been suitable as malting floors for spreading out the grain as it began to grow. The stone mortars could have been used for crushing the malt and the stone vessels for mashing the malted grain. At this early Neolithic site, they had all the necessary equipment and material culture to process the grain that they harvested into a sweet barley mash and malt liquid.

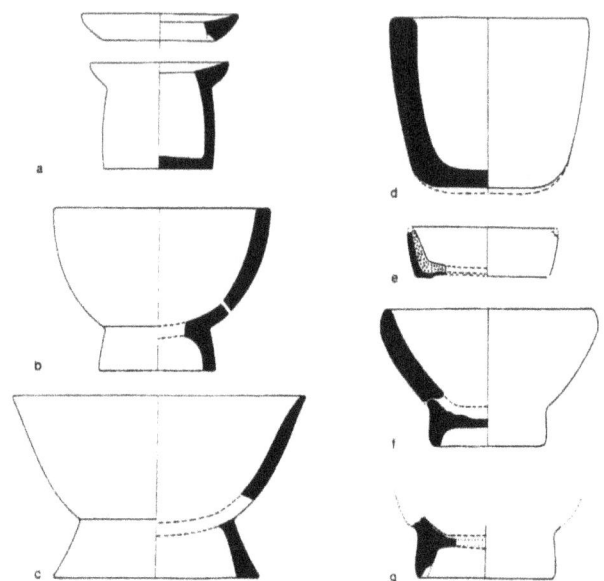

Figure 3.3

White ware vessels from pre pottery Neolithic sites in the Levant. Vessels a-d from Labwe, e-g from Tell Naba'a Faour and f from Tell Ramad (Mellaart 1975: 64).

No scale. Mellaart describes the vessels as 'large'.

Ain Ghazal, Jordan

Ain Ghazal is a large early Neolithic settlement in Jordan. Contemporary with Jericho, it is situated by a spring that flows throughout the year so it was a permanent site. Earliest radiocarbon dates obtained from Ain Ghazal are c7250 BC (Rollefson 1986:45), when the settlement was between 4 and 5 hectares in area. By c6500 BC, the site had developed to 12.5 hectares (31 acres) and was situated on both sides of the Zarqa River (Maisels 1993:92).

Excavations have revealed "architectural remains, flint tools, stone bracelet fragments, stone bowls, animal figurines, bone tools, human burials, domesticated plant remains (including wheat, fig, pea and lentil), grinding stones, ornaments, shells, plastered vessels, one obsidian tool and what have been designated 'white objects'" (Kafafi 1986:51). White Ware objects from Ain Ghazal were examined by X-ray fluorescence and were discovered to have been made of a chalky limestone material containing calcite and were carved or otherwise formed by hand. Basalt pestles and grinders/crushers were also found, as well as limestone querns and large "stone bowl" type mortars, used for crushing and grinding. Pottery technology was beginning to develop and over twenty fired sherds of vessels "from securely dated mid 7th millennium locations were located in the first four seasons" (Maisels 1993:93).

The presence of red-painted and burnished plastered floors at Ain Ghazal and of similar painted plaster floors in Jericho (Hodder 1990:35) might be evidence of a specially made malting floor but this is very difficult to prove without analyzing grains from that context for signs of partial germination. Malting floors could look like any other plastered floor surface that may be found in domestic or ritual buildings. It appears that the inhabitants of Ain Ghazal may have had adequate facilities for malting and mashing grain: they had suitable floor surfaces within substantial buildings, vessels, containers, crushing stones and ovens or kilns.

Of great interest to the excavators, indeed, a unique find, were the unusual statues and figurines that had been deliberately buried together in a pit dug into the remains of an abandoned house (Rollefson 1983, 1986, 1987, 1989). The 10 largest statues were 80-90 cm high and had been made by the construction of a reed and twig 'skeleton', bound with up with twine and covered with plaster (Rollefson 1983:32). The statues had personal details, such as noses, fingers, toes, ears and eyes and they are indicative of "sophisticated art and a ritual and social complexity" ((Rollefson 1986:45).

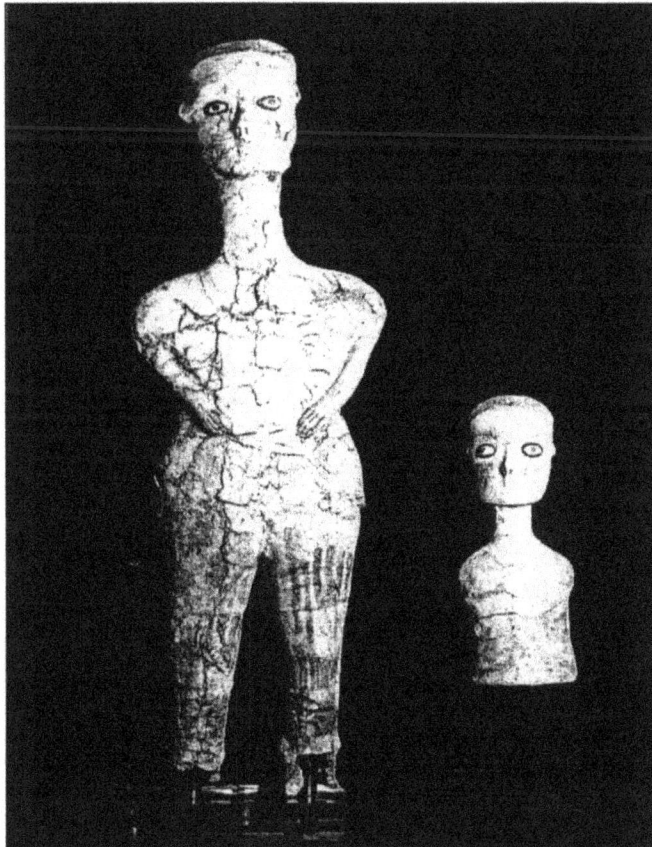

Figure 3.4
Plaster statues from Ain Ghazal, Jordan, c6500 BC. The larger figure is 84cm high. These statues indicate the complexity, sophistication and technical skills of this early Neolithic community (Collon 1995 figure 21).

These statues are striking (figure 3.4). They are described as being "quite different from the usual pregnant woman fecundity blobs, usually headless and legless, found in so many Neolithic contexts" (Maisels 1993:94). Female figurines, so disparagingly described by Maisels, are represented at Ain Ghazal and many other early Neolithic sites of the Near East, Levant and Central Europe.

Maisel's description seems to be unfair and inaccurate. Female figurines are much more than just 'fecundity blobs'. Some from Central Europe are carved from marble; others from ivory and bone and some from the 6th and 5th millennia BC have been made from baked terracotta with cereal grain impressions decorating the torso (Gimbutas 1982:154-160,204). The level of artistic and technological skill is impressive. A wide variety of beautifully manufactured figurines depicting male, female and animal entities are found on most sites of the early Neolithic throughout Europe (Gimbutas 1982). Their existence demonstrates that ritual activity and spiritual beliefs were very important aspects of life to these early agricultural people. Some rituals may have been related to the sowing, cultivation, harvesting and processing of grain and other crops.

Tell Ramad II

This site dates from the late 7th millennium BC (Maisels 1993:76). The houses at Tell Ramad were adobe huts with hearths and large basins, perhaps used for grain processing. The presence of querns, grinders, and wooden storage bins are indicative of an "agricultural regime", although hunting, fishing and fowling probably remained important subsistence activities for the people there at this time (ibid: 75).

White Ware was found at this and other contemporary sites (figure 3.2). This unbaked coil ware was made of a mixture of limestone and salty grass ashes. The mixture was soft enough to fashion and construct coiled pots and became as hard as cement when dried in the sun. Firing in a kiln was not necessary (de Contenson 1971:282,3). No complete White Ware vessels or objects have been found and the fragments so far excavated indicate that platters, deep and thick walled bowls of various sizes, pendants, cylinders and anthropomorphic figurines were being made from this material. Barley was grown from the very beginning of the settlement at Tell Ramad, being made into 'porridge', according to the excavators (ibid: 279). It is just as likely that the grain was malted and mashed using White Ware vessels and stone bowls.

4. Early pottery Neolithic cultures: 6th/5th millennia BC

Çatal Hüyük, Konya Plain, Anatolia

Çatal Hüyük is a well-known neolithic site on the Konya Plain in Anatolia. It is a large settlement tell that covers 32 acres and it was occupied between c6250 BC until c5400 BC (Mellaart 1975:98). There is the evidence at Çatal Hüyük of many features necessary for malting and mashing the grain, such as, plastered floors, ovens, kilns with domed chambers, grain bins and vessels of pottery, stone and wood (Mellaart 1967:62). Mellaart listed "… coiled baskets, with or

Figure3.5:
The terracotta figurine of a goddess, apparently giving birth and resting her hands on two leopards. It was discovered in a grain storage bin from level II at Çatal Hüyük, Turkey. The figurine is 11.8cm high.

Female figurines have been found at Haçilar, Ain Ghazal, Tell Ramad and many other Neolithic sites.

without covers, leather bags, wooden vessels and boxes with lids of various shapes, a few luxury vessels in stone and pottery" (Mellaart 1975:104). Some of the pottery vessels were deep, round based and could be described as 'bag shaped' (figure 3.6). Pottery was found in the earliest levels, with plain bowl forms predominating. Lids have not been found but Mellaart suggests that wood, textiles or leather may have been used and tied onto the lugs of vessels (ibid 1975:105). Basic equipment such as this would have provided suitable anaerobic conditions for mashing or fermentation.

Ovens and kilns are usually interpreted as being necessary for bread making and baking but they are also suitable equipment for drying malt. Mellaart interpreted the 'huge ovens', discovered in Levels IV and V, as a bakery. The grain bins, made of dried clay, were about one metre in height and were found in pairs or in rows. They were filled from the top and emptied from the bottom, via a small hole made at the base of the bin so that the oldest grain was used up first.

That the grain may have been regarded as a special, sacred and important crop is perhaps suggested by the discovery of a female figurine made of terracotta (figure 3.5). This has been interpreted as the representation of a fertility goddess and it was found within one of the grain storage bins in Level II (Davidson 1998:53). Such 'goddess' figurines are

found in at a great many other sites in Anatolia, for example, at Haçilar.

The symbolism within the ritual areas at Çatal Hüyük is complex and powerful, with women appearing to have control over wild creatures, such as leopards and lions (Hodder 1990:5ff). The images on the walls could depict hunting deities. The female deities represented as figurines at Çatal Hüyük and at other similarly dated sites are powerful images (figure 3.5). Hilda Ellis Davidson sees them as embodying reproductive power, fertility and the domestication of the wild (Davidson 1998:53,138). She notes that, from later prehistory and documented in early history, there are many powerful female deities strongly associated with fertility and with the sowing, cultivation and processing of grain crops. For example, in Egypt, Isis was the deity that presided over bread, beer, green wheat, barley and also flax. She was reputed to have taught the people how to grow their crops and make their bread and beer.

The connection between bread and ale is emphasised in a 'Hymn to Hathor', another Egyptian Goddess who is sometimes identified with Isis. She is addressed as "Mistress of Bread, who made Beer." Demeter, the Greek goddess, was associated with the sowing of grain and her principal festival took place before the first ploughing of the season.

Figure3.6
Wooden (5-8), stone (4) and pottery (1-3) vessels from Çatal Hüyük, Anatolia, from levels II, III, IV. The pottery vessels appear to be suitable for processing barley mash and liquids. The function of the stone object (4) is unclear. The wooden vessels are suitable for the consumption of food and drink (after Clarke, J.G.D. 1977: figure 27,p71)).
No scale.

Animal sacrifice and ritual activities were associated with the ploughing of the land, the sowing of grain and the harvest and these rituals are connected with the cult of Demeter (ibid:54*ff*). It is probable that rituals such as these had their origins in the early Neolithic and they are associated with the earliest cultivators. The archaeological evidence of Çatal Hüyük, where sacred and ritual spaces, images of deities and elaborate wall paintings have been uncovered (Mellaart 1967:77) seems to support this idea.

Ceremony and ritual activity were clearly very important to the early Neolithic people on the Konya Plain. It seems very likely that malt and ale were being made at this time, perhaps ritually and as a special product. The association of a powerful female deity with grain storage is therefore a credible concept in this early Neolithic context.

5. Fermentation of barley wort into ale: 4th/3rd millennia BC

Godin Tepe, Zagros Mountains

The earliest chemical evidence so far discovered for the manufacture of ale has been found at the site of Godin Tepe, an Uruk site of the late 4th millennium BC. The settlement appears to have been a trading centre and it is located in the Zagros Mountains of modern Iran (Joffe 1998:303).

An organic residue inside a double handled jug, found in Room 18 of the Oval Enclosure on the citadel of Godin Tepe, was analysed and identified as 'beerstone' or calcium oxalate. This is a substance that always settles out on the internal surfaces of vessels that are regularly used for the fermentation or storage of barley wort (McGovern 1992).

Residues of tartaric acid were found in other vessels that had been stoppered and were found lying on their sides in the same room, indicating that they probably had contained wine rather than ale. According to one of the excavators, Virginia Badler, the presence of wine and beer containers in such an important area of the enclave indicates that these products were being manufactured and distributed along with other foodstuffs and possibly transported to and traded with people in the lowlands (Badler, quoted in Joffe 1998:304). Joffe points out that the discovery of another site with similar ceramic assemblages and architectural structures at Habuba Kabira South on the bend of the Euphrates perhaps indicates "parallel redistributive mechanisms" (Joffe 1998:304).

Brewing in Mesopotamia

By the late 4th/early 3rd millennia BC malting, mashing and fermentation were being practised on a large scale in Mesopotamia. Approximately 40% of the barley harvested was being used for malting and brewing purposes (Corran 1975:15). Cylinder seals, evidence of the highly organised administrative system of the Uruk culture, have been found with illustrations depicting the consumption of an alcoholic drink, presumably either wine or ale, being drunk from large vessels through straws (figure 3.7).

Evidence for the large-scale production of bread, malt and ale has been found at the 3rd millennium site of Abu Salabikh in southern Iraq, a large settlement complex. So far 29 fire installations from Area E have been excavated, including several 'tannurs' or bread ovens, clay lined open hearths and one example of a two storey oven (Crawford 1981:105) which could probably have functioned for drying malt. Further research clearly needs to be done into the differences in kiln or oven design and their specific function.

There is evidence for the centralised storage of grain in specially constructed buildings during the 3rd millennium BC in Northern Mesopotamia. A number of sites with storage and processing facilities for amounts of grain in quantities "well beyond that needed to support the local population" (Irving 1999:43) have been located and excavated in the middle Khabur Valley, northeast Syria (figure 3.2). These sites, namely Tell Atij, Kerma, Tell al-Raqa'i, Bderi and Tell

Ziyadeh, are clustered along the banks of the Khabur River and are dated to the mid to late Ninevite V period.

At Tell Atij six semi-vaulted structures with plastered interiors were interpreted as grain silos. Storage jars were plaster lined, perhaps to aid the preservation of the contents. A suggestion that can be made in the light of experiments described in Chapter One is that this plaster lining of pottery jars was to facilitate the containment of a liquid product, such as wort or ale. At Tell al-Raqa'i the existence of silos, mud brick platforms, ovens and pottery sherds suggests processing of the grain, probably by malting and kilning.

The dominant grain grown in this region was barley. It may have been grown as an animal fodder crop to feed the herds of the rich and powerful elite groups (Irving 1999:44), but it is also the fundamental ingredient for the manufacture of sweet malts and ale. It has been suggested that the extent of the agricultural and administrative development in the middle Khabur Valley indicates increased social complexity and the rise of elite and powerful individuals or groups who had control over the trade and distribution of agricultural produce (ibid). Clearly, the Khabur Valley was an important trade route for the exchange of goods and products destined for both the south and the north using river transport. One of these products may have been ale.

Both the manufacture and the consumption of ale and wine were of significant social, religious and economic importance to Sumerian culture. Alexander Joffe (1998) sees the manufacture, exchange and consumption of alcohol as forming "a significant element ...in the emergence of complex, hierarchically organised societies." His detailed analysis, based on Michael Dietler's model (1990), concentrates on the social, economic and political aspects of the consumption of alcohol rather than on the practicalities of its manufacture.

From the available evidence it seems that beer and wine were being manufactured and consumed on a huge scale during the Ubaid and Uruk periods. Pottery was being mass-produced and large public buildings were being constructed, both administrative and religious (ibid:303). Alcoholic beverages had become an important and integral part of the economic and religious systems, with complex redistribution systems and centres of mass production. Numerous administrative texts record brewing as a craft activity and the redistribution and trade in beer. Drinking scenes appear on seals, often in ceremonial or celebratory situations. The Royal Standard of Ur, dated to c2600 BC and found in grave number 779 of the Royal Cemetery, depicts feasting and celebrations after a military victory. Animals are depicted being brought either for slaughter or as booty and grain is being carried, apparently as important and valuable a commodity as the animals were.

The Sumerians brewed eight types of ale from barley, eight from wheat and three from mixed types of grain (Vencl 1994:308). Both the Sumerians and the Egyptians mashed the malted barley by making so-called 'beer bread' or *bappir*. Corran (1975:17) argues that "with the making of beer bread, the processes of baking and brewing have become linked together". According to annotations referring to *bappir* and ale in the Sumerian and Akkadian dictionaries, *bappir* was, however, only eaten during times of famine or necessity

(Katz & Maytag 1991:27). The grain was being mashed as *bappir* and then stored as a raw material for the purpose of making ale not as bread for daily consumption. Certainly, by the early 3rd Millennium BC, a 'drinking culture' was well established in the Near East and Levant.

In 1990 Solomon Katz and Fritz Maytag reproduced a Sumerian-style ale based on the methods described in the Hymn to Ninkasi and working in association with the Anchor Brewing Company of San Francisco. A few modifications were made for legal reasons and also because they were unsure that the mix would contain enough sugars for fermentation. Honey was added instead of dates to the *bappir* and malt extract was added to the mash tank. Nevertheless, in spite of these changes, they proved that the ancient techniques worked and the resultant ale was described as having "the smoothness and effervescence of champagne and a slight aroma of dates" (Katz & Maytag 1991:33).

The mashing experiments undertaken as part of this research indicate that their fears of inadequate amounts of grain sugars for fermentation were completely unfounded. With the correct equipment, conditions and temperatures and the specialised knowledge of the crafts of malting and mashing it is possible to make a wort with a very high sugar content from the barley grain alone.

Figure 3.7
Early Dynastic seal impressions depicting drinking scenes. Straws, presumably made from reeds, are used to avoid drinking the chaff and sediments (after Singer 1980, Amiet 1980).

6. Brewing in Egypt: 3rd/2nd millennia BC

Ale was a staple item of diet in Egypt from the 3rd millennium BC onwards, a drink for both the poor and the wealthy. It was drunk on a daily basis and it was also brewed for state occasions and festivals, being an essential part of offerings to gods and goddesses (Lucas 1962:12-16, Davidson 1998:138). Ale was of considerable economic and social importance. Frequently it was a part of wages or payment for work done or services rendered. The villages of Amarna and Deir el Medina provide excellent evidence for large-scale breadmaking and brewing during the 2nd millennium BC (Samuels 1996).

Due to the dry climate of Egypt organic residues are perfectly preserved on pottery. Budding yeast cells and even signs of enzyme attack on starch granules during germination can be seen using scanning electron microscopes (Samuel 1995). Residues on Eighteenth Dynasty Egyptian pottery were first noticed in 1928 by Dr J. Gruss, who observed starch grains, yeast cells, moulds and bacteria under a microscope and concluded that this yeast was either deliberately or accidentally cultivated by the Egyptians (Lucas 1962:16).

Techniques of grain processing are illustrated in scenes that decorate Egyptian tombs (see figure 1.8). Statuettes and models were sometimes placed in tombs in order to re-create all necessary provisions for the deceased in the afterlife (Samuel 1996:3). Delwen Samuel of Cambridge University has recently tested these ancient grain processing techniques in collaboration with the Egypt Exploration Society and Scottish and Newcastle Breweries. A re-created Egyptian beer was made and it was aptly named 'Tutenkhamun' (Samuel 1996). It was, apparently, sweet and good to drink.

If the manufacture and consumption of ale was of such great importance to Sumerian and Egyptian Bronze Age cultures, as is suggested by Joffe (1998, then what of its significance in a Neolithic and European context? The next three chapters evaluate the possibilities for the processing of barley grain into sweet malts and ale in Neolithic Europe, during the Orcadian Neolithic and in the Neolithic British Isles. As it has been so far, the emphasis is on the assessment of the practical aspects of the material culture, buildings and equipment necessary for this kind of activity (see figure 1.15).

The consumption of ale has been the subject of several archaeological and anthropological studies (Burgess and Shennan 1976; Dietler 1989,1996; Hayden 1996; Joffe 1998). The intention of the next three chapters is to concentrate on assessing the extant archaeological evidence for the ritual, perhaps even magical activity of the manufacture of malt, malt sugars and ale from the grain.

CHAPTER FOUR
Grain in Neolithic Europe.

1. The transition to agriculture in Europe

The introduction of grain cultivation and processing into Europe is an enormous and complex area for investigation. Much has been written about the particular reasons for the Mesolithic/Neolithic transition in Europe. There are many theories and assessments of the reasons for the transition from a hunting, gathering and fishing lifestyle to an agriculturally based one. Archaeological debate has been, for the most part, theoretical, and has used anthropological and ethnographic analysis to reconstruct the past (Tilley 1996:25). This chapter will investigate practical aspects of early Neolithic cultures in Europe and assess the viability and potential of their material culture for malt and ale production.

Ian Hodder (1990) proposed the concept of 'domus' which focuses upon the importance of hearth and home and upon the control or domestication of the wild in the early Neolithic. The definition of 'domus' is best summarised in his own words as "practical activities carried out in the house, food preparation and the sustaining of life. But it is also an abstract term. Secondary, symbolic connotations are given to the practical activities, leading to the house as a focus for symbolic elaboration and the use of the house as a metaphor for social and economic strategies and relations of power" (Hodder 1990:44).

Christopher Tilley (1996) has interpreted the Neolithic in ethnographic terms. He refers to food as "a highly symbolic medium. Its production, consumption and distribution is never just a practical and technological matter but is loaded with symbolism and hedged around by political and social relations. Acts of eating, sharing, food preparation and consumption must, in the Neolithic as in our own and other societies have formed a basic medium for sociability and for ritual practice" (Tilley 1996:111). The contribution made by women to these fundamental and important areas of life, such as food production, nurturing babies and children and other domestic activities is significant.

It seems that, at last, the focus of archaeological thought and interpretation is turning away from concepts of death and an obsession with burial rites and rituals. Archaeological thought is now turning towards more practical aspects such as food production, food processing, consumption and the rituals of daily life.

The Neolithic involves the introduction of new technologies such as the introduction of ceramics technology, which transformed food preparation methods and techniques. At the same time, grain products were an important and new addition to the diet in early Neolithic Europe. As well as providing carbohydrates in the form of flour and bread, grain is the primary ingredient for malt and ale. As has been discussed in earlier chapters, ale was an important commodity in ritual, social and economic terms in the context of Mesopotamian, Sumerian and Egyptian cultures of the Neolithic and Bronze Age. Alcoholic drinks that produced intoxicated states were significant aspects of ritual belief systems, ritual practice and ritual behaviour in these early complex societies. For example, the Sumerians had a goddess of brewing named Ninkasi.

As well as having ritual significance, alcoholic drinks were also an important part of the social and political economy, being paid as a tribute to the elite and powerful as well as being part of a 'wages' system in the construction of large monuments and tombs. Competitive feasting between rival elite and powerful groups within society also seems to have occurred (Joffe 1998:297). The concept of 'competitive feasting' has recently been introduced to the discussion of cultural evolution in Europe. Brian Hayden suggests that "one of the best candidates for a prehistoric competitive feasting system is the European Neolithic" (Hayden 1996:141). There were feasting sites at causewayed enclosures and at megalithic tomb sites, as well as evidence for complex trading systems, exchange networks and high status individuals within neolithic society. Special ceramic vessels were used at feasts (figures 4.9, 4.10, 4.11). It is very likely that special foods and drinks, for example ale or mead, were consumed at these and at other ritual or social occasions.

Not all feasts need to be competitive or ritualistic in nature. Feasting can be a celebratory event, creating social bonds between individuals and groups. It can involve 'work-party feasts', that is, feasting as a payment or reward for the collective accomplishment of a task by the community. Monumental constructions involving communal effort are an important aspect of the early and later Neolithic in Europe and the British Isles. Such work may have been rewarded by the provision of a feast to the workforce (Dietler 1996:104).

In the light of the evidence presented so far in this thesis, it is highly improbable that early and later Neolithic groups of Europe were cultivating barley and wheat only for their porridge or bread-making potential. It is even more unlikely that these were the special foods that people consumed at feasts. It has been proposed that ale did not become a part of life in Europe and Britain until the Bronze Age (Burgess and Shennan 1976). Andrew Sherratt has suggested that "pure beers did not appear until the Iron Age" although he acknowledges that stimulants of other kinds, such as cannabis, henbane and opium poppy seeds, may have been a part of Neolithic ritual activity (Sherratt 1995:25). Alcohol is a drug that alters the mood and consciousness of people. It is also a stimulant. It is easy to manufacture, provided one has knowledge of the special processes necessary.

The intention of this chapter is to show that early Neolithic groups of Europe had the necessary and required material culture to convert the wheat and barley grain that they grew into malt and ale, based upon the experimental grain processing as described in Chapter One. The proposal is that the manufacture of malt and ale was a major driving force behind the change from a hunting and gathering lifestyle to that of grain cultivation within Europe. The attraction of grain cultivation lay in a desire for the products that could be manufactured from the grain rather than in the acquisition of a new lifestyle of farming.

2. European Neolithic: 6th /5th millennia BC

The 6th and 5th millennia were a transitional period in Europe, when the lifestyle of the indigenous Mesolithic population was influenced and changed by new ideas of grain processing. Cereals are not a native European plant. They were introduced from the Near East and Anatolia, where they grow in the wild (Zohary & Hopf 1988:13).

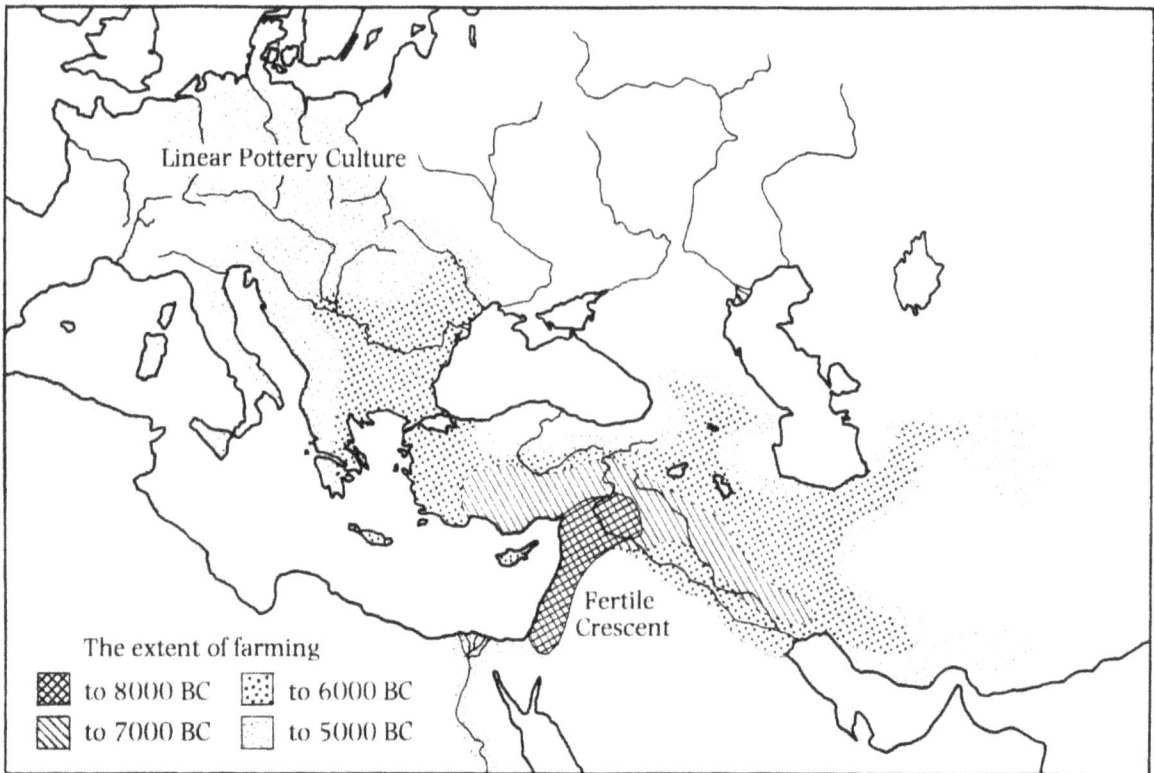

Figure 4.1
Map that shows the spread of agriculture from the Fertile Crescent into Central and Northern Europe, based on radiocarbon dates (Sherratt 1980).

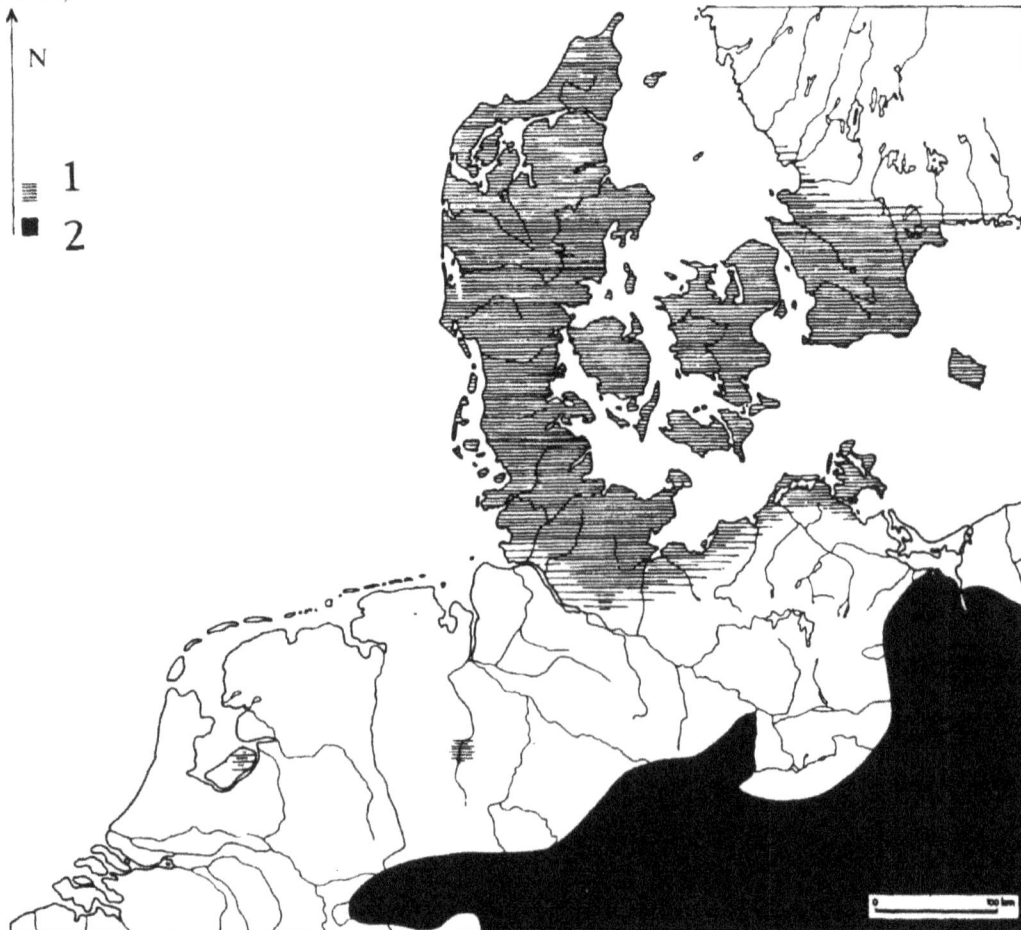

Figure4.2
Map that shows the areas inhabited by the Mesolithic Ertebølle/Ellerbeck groups (1) and the Linearbandkeramik early agricultural groups (2) in the 5th and 6th millennia BC (Tilley 1996).

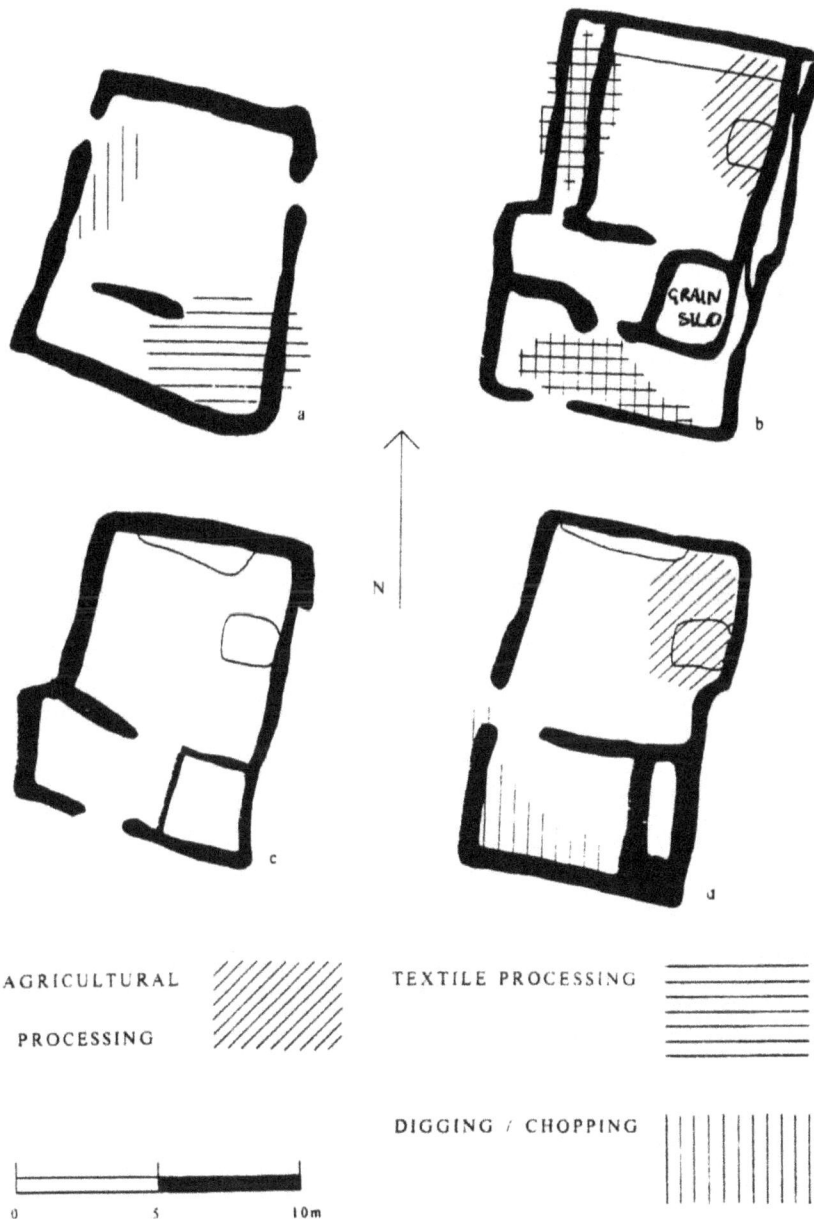

Figure 4.3
Ground plans of House 59 at Tell Ovchorovo in northeast Bulgaria, showing changes made to the house over the years and the areas for the different grain processing activities (after Bailey 1996: 151).

(a) The first house had no hearth, no clay platform and two work areas.

(b) After 40 years the house was rebuilt with room divisions, perhaps even a second storey, a hearth by the east wall, a clay platform against the north wall and a grain silo. This house burned down.

(c) The house was rebuilt with two rooms, a grain silo, a hearth and a clay platform.

(d) Grain processing continued around the hearth area where many grinding stones and pottery vessels with volumes between 500cc and 10,500 cc were found.

AGRICULTURAL PROCESSING

TEXTILE PROCESSING

DIGGING / CHOPPING

0   5   10m

In Europe, there was not a single 'Mesolithic' or 'Neolithic' way of life, but rather "a diverse set of lifestyles and practices" (Tilley 1996:70). Some Mesolithic groups, for example the Ertebølle/Ellerbeck cultures of the Northern European coast, maintained their hunting and gathering lifestyles for over a millennium, living alongside well established agricultural communities, namely, the Linearbandkeramik culture who lived along the river valleys of central Europe (figures 4.1,4.2).

Northeast Bulgaria: Tell Ovchorovo.

After the grain has been sown, cultivated, harvested and stored it has to be processed for consumption. During the experiments described in Chapter One it quickly became apparent that this is hard, skilled and labour-intensive work. Water and wood have to be fetched. Heavy storage or cooking pots need to be lifted and equipment has to be washed and cleaned on a regular basis to avoid contamination. It is also important to maintain a careful and close watch on the grain in all stages. Even making relatively

small amounts of malt or ale is a task more easily achieved by a group of people rather than by an individual.

Evidence for collective grain processing has been found in Northeast Bulgaria at Tell Ovchorovo, a settlement tell almost 5m high and 100m in diameter, situated about 100 km west from the shores of the Black Sea. Radiocarbon dates from the early 5th millennium BC have been obtained (Bailey 1997:142,149). There are many such settlement tells in Bulgaria but only a few have been excavated so far. Those that have been excavated reveal the remains of complex dwellings. Some may have been two-storeyed buildings with multiple entrances and may have had up to 5 rooms that were "full of furniture, conveying a sense of comfort and domesticity" (Chapman 1991:158). Most of the buildings seem to have been dwellings but in some "figurines and cult vessels were found which were used in some form of domestic ritual" (ibid:158). Ritual items associated with this culture include anthropomorphic figurines of males, females and animals, artefacts inscribed with ritual designs and decorated pottery vessels. This was an early agricultural society with a complex social and ritual structure that appears to have been, in many ways, similar to early Neolithic pottery cultures of the Near East.

House 59 at Tell Ovchorovo is one of 112 houses on the settlement tell. It was excavated between 1971 and 1973 (Todorova 1976a, Todorova et al 1983). Two principal work areas were defined within the house (figure 4.3). One area was for textile producing activities and the other was for grain preparation and processing (Bailey 1996:144). As well as a large grain silo that was situated between the main room and the doorway there were grinding stones, pottery sherds, an oven, a hearth and an interesting raised clay platform. Grain processing was an increasingly important activity over time within the settlement and Douglas Bailey (1996) has interpreted the house as "a work station for the storage, parching and grinding" of cereal grains.

What exactly is meant by this description of grain processing? If the grain was made wet before being 'parched' and then 'ground' then it may have begun to germinate and, if so, the necessary enzymes that convert starch into sugars will have been activated within each grain. The necessary morphological and physiological changes for malting have occurred. Bailey notes "redistribution may have occurred on a settlement wide basis with many different individuals bringing their small quantities of grain to be processed in the house before storing the grain in their own houses" (ibid: 150). Given the limited number of products that can be made from barley or wheat, namely flour, bread, porridge, malted grain or ale, what is the evidence for malt and ale manufacture in House 59? Pottery sherds were found with potential volumes of vessels ranging from 500cc (about a pint) in early levels to 10,500cc (2 gallons) in the later levels of House 59. Most of the pots were less than a litre in capacity and would have been suitable as vessels for consumption. There was only one very large vessel, 10,500 cc in volume. This pottery assemblage is consistent with what might be expected for the manufacture of ale. One single large vessel is all that is needed for the fermentation of a sweet wort and many smaller vessels are useful for consumption of the end product.

Bailey concludes that the house probably had a special function as a grain processing centre for the community rather than functioning as a house or dwelling. This kind of behaviour and activity compares well with the central grain storage and processing centres that are evident in the Khabur Valley during the 3rd millennium BC, as described in the previous chapter. The process described by Bailey as 'parching' was probably the drying out or kilning of partially germinated grain (malt). Some of the pottery vessels were quite large and, although there is no organic residue evidence yet available to confirm that they were used for storage and fermentation of barley wort, I believe that this is one of their most likely functions.

The Linearbandkeramik

There is widespread evidence of mixed agriculture during the 6th and 5th millennia BC in Europe. Sites of the Linearbandkeramik, a remarkably homogenous cultural group who constructed large, elaborate longhouses, grew several varieties of crops, kept cattle, sheep and pigs and who also made fine decorated pottery (figure 4.4) have been found along the fertile river valleys of the Danube, the Rhine and other significant rivers and their tributaries throughout central Europe (Ehrenburg 1992:90). The Linearbandkeramik culture flourished between c5500 BC and 4800 BC. Prestige artefacts, such as spondylus shells from Greece (Halstead 1993: 607) and axes of basalt and of amphibolite, have been found at Linearbandkeramik sites and such finds indicate long distance travel, communications, trade and exchange between these early agriculturalists and other cultures (Thorpe 1996:33). The knowledge of grain processing techniques would have been passed on through these far-reaching trade and exchange networks.

One view of their culture is expressed by Peter Bogucki (1988). He proposes that they were primarily cattle herders, one of the main reasons for this lifestyle being the production of milk and meat. Small vessels with narrow spouts have been found in childrens' graves of this period that might perhaps indicate the feeding of cow's milk to infants (Vencl 1994:302). The processing of milk products is perhaps indicated by the ceramic sieve pottery (figure 4.4). Sherds of this kind of pottery, generally interpreted as being the remains of sieves, presses or strainers of some kind, have been found at many Linearbandkeramik sites (Bogucki 1988:89).

As well as being herders, the Linearbandkeramik also cultivated a wide range of crops including pulses and grain such as einkhorn, emmer wheat and barley for processing and consumption. Both pulses and cereal grains require abundant water for their preparation. The finely painted and decorated large pottery bowls and small drinking cups are generally understood to show "the importance of consuming food and drink in the context of feasting" (Thorpe 1996:32). This finely made pottery may have been used for the manufacture and consumption of a prestigious liquid, probably ale. Carbonised grains are frequently found at these early Neolithic sites (Bogucki 1988). They could perhaps indicate the accidental over kilning of malt or they may indicate the accidental burning of the wooden building used to make, store or process malt.

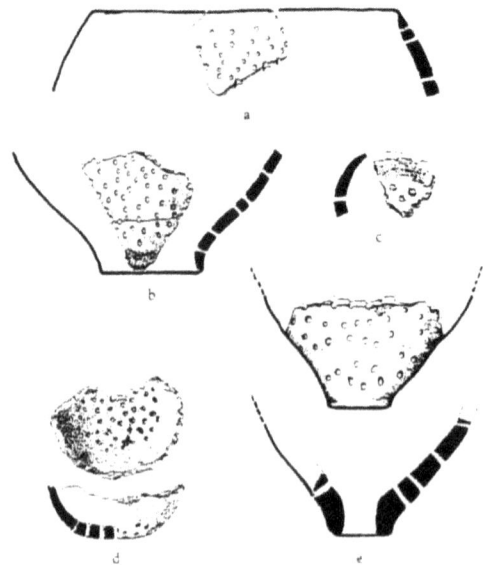

Figure 4.4
Ceramic 'sieve' sherds from Linearbandkeramik sites in Central Europe. a, b, & e from Brzesc Kujawski, Poland; c from Murr, Germany; d from Ditzingen-Schockingen, Germany. (Bogucki 1988). No scale.

Figure 4.5
Linearbandkeramik pottery bowl and cups, flint knives, arrowheads and polished stone adzes. The pottery is finely made, decorated with incised lines. It was used at ceremonial feasts, perhaps for containing and consuming ale. These pots are in collections at Rheinisches Landesmuseum, Bonn. (Ehrenburg 1992).

In my view, it is a very strong possibility that some of the grain processing taking place within this early Neolithic culture was the manufacture of malt and ale. Water would have been accessed from local springs, streams and rivers or even specially constructed wells. Several wells dating to the mid 6th millennium BC have been found at a number of Linearbandkeramik sites, for example, at Mohelnice, Most, Remsdorf, Zipfendorf and Erkelenz-Kuckhoven (Vencl 1994:300). Weeds and wheat chaff representing grain processing waste have been found in rubbish pits at a number of settlement sites (Thorpe 1996:35). The cultivation of grain and the manufacture of malt and ale seem to have been an important element in their subsistence and lifestyle.

The fine elaborate longhouses that they built appear to have been used as dwellings and shelter for both humans and animals. They may also have functioned as grain storage and processing centres. These longhouses were the focus of Linearbandkeramik society and there was considerable variety in both size and in design.

Some longhouses were up to 8 metres in width and up to 45 metres in length, others were much smaller buildings (Thorpe 1996:34). Unfortunately, floor surfaces have not been preserved since many of these settlements or villages are situated on loess soils, good agricultural land that has been heavily ploughed in modern times. This adds enormously to the problem of interpreting this cultural group, since hearth and floor evidence have been completely destroyed. Posthole evidence, however, indicates the internal division of the houses into distinct areas where different activities took place. It is likely that there were significant functional differences between the buildings.

Some of these structures can be interpreted as "clubhouses" or as "ceremonial houses" (Soudsky in Thorpe 1996:35). It is also possible that "primary processing" of the grain was carried out in some of the longhouses. Malting, a primary process in the conversion of grain into malts and ale was probably the activity concerned here. Modderman proposes that the grain distribution and control may have been in the hands of a priest or other community leader who controlled a central granary for the village (in Mitford ed, p275). Central control of grain might reflect its importance and value as a crop or it may simply have been more practical to store the grain collectively.

Anick Coudart (1987,1991) has interpreted the Linearbandkeramik as having been a largely egalitarian society but with elements of status differentiation of individuals within the group as a whole. She interprets the grain cultivation and its subsequent storage and processing as being a collective activity among extended family groups. Ehrenburg (1992) points out that in Linearbandkeramik burials querns are found in association with only female burials and not with males. She concludes that women "would almost certainly have been responsible for most, if not all, the agricultural work" (Ehrenburg 1992:90). This would include the cooking of food and the preparation and processing of grain as well as tending crops in the field and harvesting them.

There are many problems in the interpretation of the Linearbandkeramik culture not the least of which is the lack of well-preserved floor and hearth evidence. This makes a complete understanding of this early Neolithic culture difficult. In spite of this difficulty it seems clear that they had an adequate material culture and the necessary requirements for the processing of grain by malting, mashing and fermentation. These activities were very likely to have been carried out by women in the community who were the primary crop producers and processors.

3. Late Mesolithic and Early Neolithic: 5th/4th millennia BC

Ertebølle and Ellerbeck cultures: N. Europe, S. Sweden:

At the same time that the early agricultural communities described above became established there were other European groups that maintained the Mesolithic lifestyle. Fishing, gathering and hunting were their main subsistence activities. These groups, whose settlement remains have been discovered along the coastline of Denmark, in southern Sweden and in northern Germany as well as by several inland lakes, are referred to as the Ertebølle or Ellerbeck cultures and they are believed to have lived in permanent settled communities (Bogucki 1988:108).

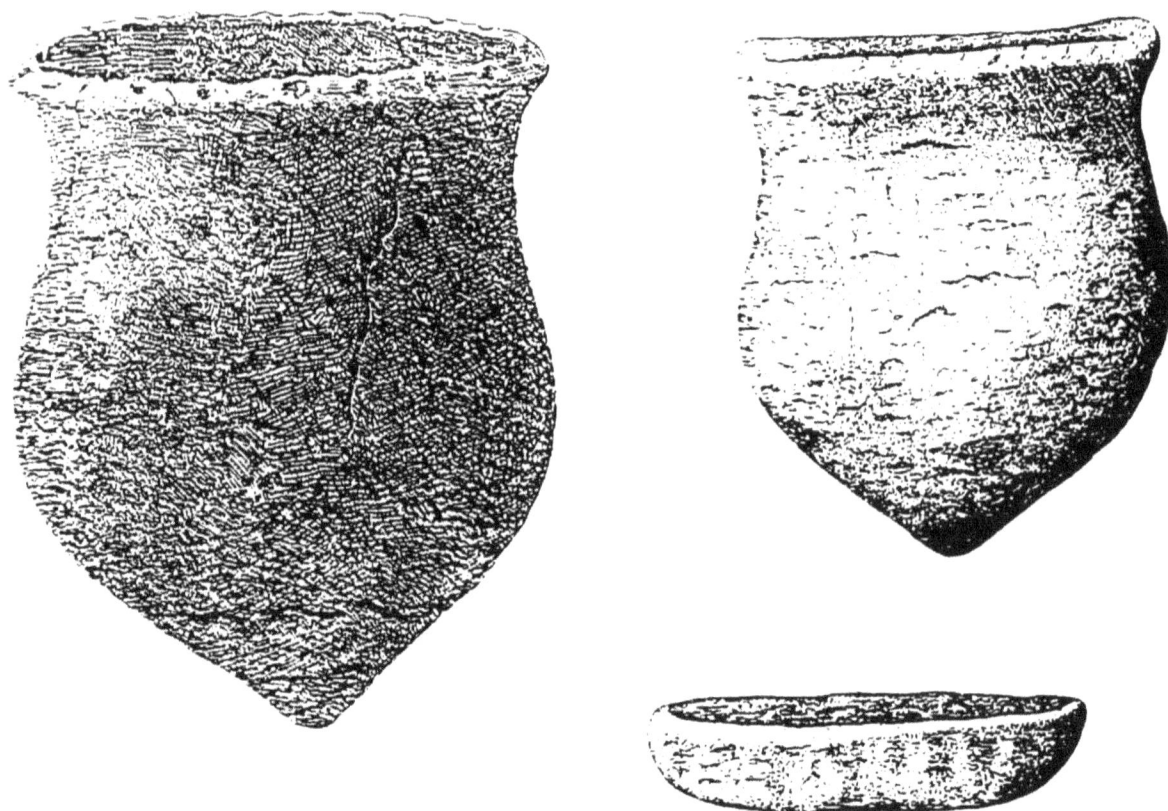

Figure 4.6
Characteristic forms of Ertebølle pottery. The containers have pointed bases and are not drawn to scale (left from Muller 1918 and right Burenhult 1982). The shallow bowl is a blubber lamp (after Tilley 1996 figs 1.14,1.31)

Anders Fischer (1981) investigated the distribution of Danubian shaft hole axes made by Linearbandkeramik groups. He believes that contact and trade must have occurred since some of these axes have also been found in late Ertebølle contexts. Therefore, trade in perishable goods such as furs and foodstuffs probably also occurred between these groups. The Mesolithic communities continued their fishing, hunting and gathering lifestyle for almost a millennium alongside the well established Northern Linearbandkeramik groups (figure 4.2). The Ertebølle and Ellerbeck cultures "were not starving foragers waiting to be enlightened by the appearance of food production" (Bogucki 1987:48). They were complex social groups living successfully by fishing, foraging and hunting, as well as by trading with other groups.

When the transition to agriculture occurred it was a swift process, taking place over just a few generations (Blankholm 1987:156). Reasons for this sudden transition are still unclear and they are much debated. Rowley Conwy's (1984) model of the eventual acceptance of agriculture because of a seasonal decline in marine resources, specifically the oyster, is questioned by Blankholm. He points out that only the coastal groups were dependent upon the oyster as a food resource and he concludes, "there is still a long way to go before we may adequately outline and explain the transition to farming in Southern Scandinavia." (Blankholm 1987:161).

Troels Smith (1967) believes the Ertebølle may have been 'semi agrarian' but this is not an idea that has received much support until recently. A single grain of cereal pollen type was found in a level dated to 5200 BC at Farups Moss,

an Ertebølle site in the Ystad area of Scania (Thorpe 1996:75). This is minimal evidence from which to interpret an agricultural lifestyle for the Ertebølle. At Trundholm, in North West Zealand (Kolstrop 1988), pre elm decline wheat-type pollen, the pollen of plantago lanceolata (a weed often associated with cereal cultivation) and a barley or large grass type of grain have been found. These finds can be interpreted in several ways. There may have been a "consistent or low level presence of cereal cultivation in the later Ertebølle" (Thorpe1996: 75) or the pollen may have been naturally transported there, perhaps by the wind. A final possibility that cannot be ignored is that these finds are grass pollen and they have been misidentified.

Ertebølle pottery first appears in the archaeological record around 4600 BC (Thorpe 1996: 71). Small dishes that are thought to have been used as blubber lamps and large storage or cooking jars with pointed bases have been found (figure 4.6). The analysis of residues of charred food remains on this pottery has produced some extremely interesting results. Not surprisingly, herring and cod bones as well as fish scales have been identified. Significant quantities of land-based ingredients, presumably fruits, vegetables and seeds, were being collected and cooked.

The most interesting residue analysis was one of the vessels from Tybrind Vig that appeared to have contained "a fermented porridge with ingredients that included hazelnuts and possibly blood (Arrhenius & Liden 1988)" (Thorpe 1996: 71). The substance that was analysed may be the remains of the sludgy residue that accumulates at the bottom of vessels used to ferment wort. Porridge is a starchy carbohydrate that

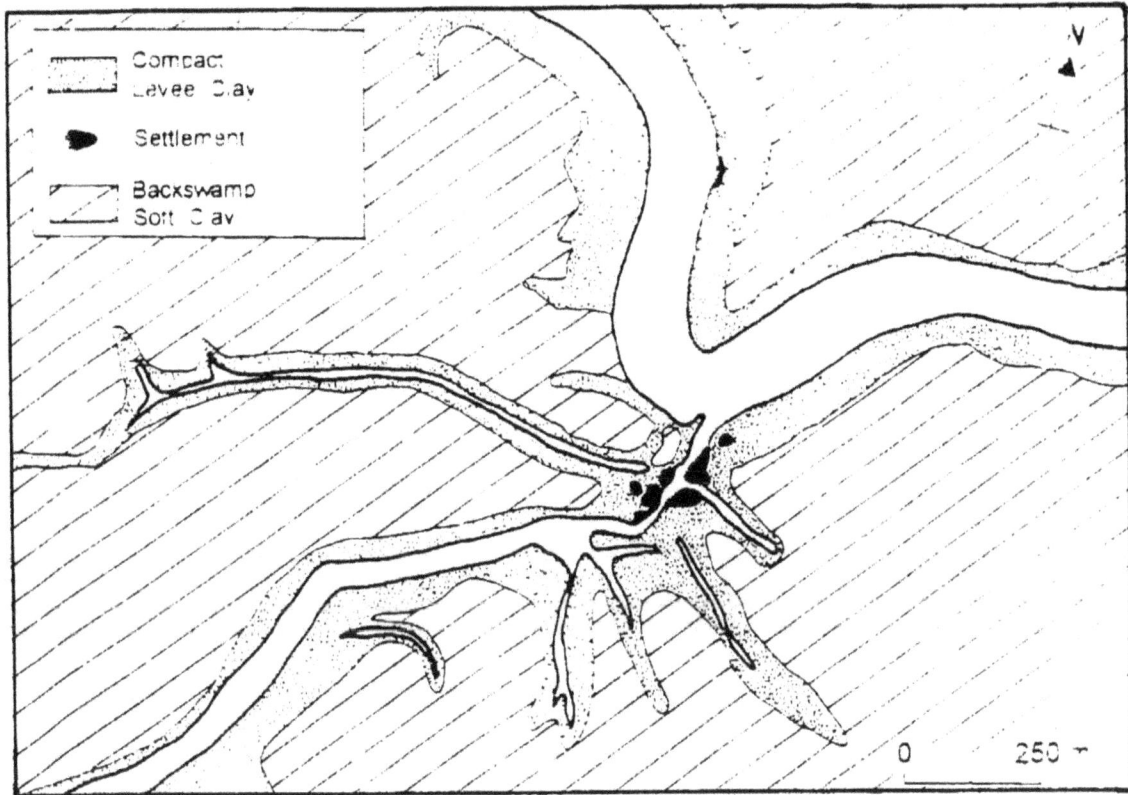

Figure4.7
The location of the sites of the Swifterbant group, at the confluence of the Rhine and Maas rivers, Central Europe (after Barker 1985, Thorpe 1996:53)

cannot be fermented (see Chapter One). The blood is an interesting interpretation and the residues need to be further analysed for clarification. A similar interesting analysis of charred residues has been described as a "highly nutritious porridge consisting of mixed wild seeds, hazelnuts, egg-white and possibly blood that had been allowed to ferment (Arrhennius 1984)" (Tilley 1996:25). This was found on a pottery vessel from the Ertebølle site at Loddesburg, western Skane. Again, this residue probably represents brewing sludge or sediments. The identification of 'egg white' or 'albumen' within the mixture is particularly interesting. These are by-products of mashing, the four major soluble proteins of barley products being albumen, globulin, gliadin and lutelin of which the first two predominate (Line 1980: 129).

These interesting analyses of residues from Ertebølle pottery with similar mixtures of processed grain indicate that the Ertebølle could have been mashing or fermenting grain. If so, they appear to have been making very unusual and extremely interesting brews. The apparent addition of 'blood' and hazelnuts is intriguing. Without access to the details, the original excavation report and residue analyses, it is impossible to comment further on these finds, except to note that albumen is one of the primary soluble proteins in a barley mash. It is clear that much further research into these interesting residues and into the lifestyle of the Ertebølle is required.

How did the Ertebølle obtain the grain? Did they cultivate it, did they obtain it by gift exchange or by trading? It is impossible to be sure. Jennbert (1988, quoted in Thorpe 1996:91) believes that the grain may have been imported, either as whole grain to be processed or as the finished

product, namely, as flour, porridge, and bread or as manufactured alcoholic drinks. The residues found on Ertebølle pots seems to suggest the importation, trade or exchange of the whole grain from nearby agricultural groups and the subsequent processing of it into an interesting kind of fermented drink by the Ertebølle themselves.

Swifterbant and Hazendonk; Ijsel, Holland
The confluence of the Rhine and Maas rivers on the North European plain forms an unusual estuarine environment, with lowland forest and river resources but also extensive marshes, tidal flats and peat bogs. Here there is good preservation of plant material at Mesolithic/early Neolithic sites in the area (Bogucki 1988: 158). Over 50 sites have so far been located in the Swifterbant sand dunes area (figure 4.7). Early cultures were aceramic until c4700 BC when S-profile pots with pointed bases, similar to those made by the Ertebølle groups, begin to appear in the archaeological record (Thorpe 1996:53). At later sites dating to c4300 BC onwards the faunal assemblage includes the bones of domesticated cattle and pigs as well as wild birds, game and fish, indicating a change to a mixed agricultural lifestyle. Plant remains were abundant, having been well preserved in the wet environment, and there is evidence of apples, berries, naked barley and wheat as being part of their diet. Cereal chaff, including internodes as well as grain, have been found, indicating the local cultivation of the grain and, therefore, the local processing and consumption of its products.

Further inland, at Hazendonk, at sites dating to c4200 BC domesticated cattle were being kept. These sites may have been "intermittently inhabited and surrounded

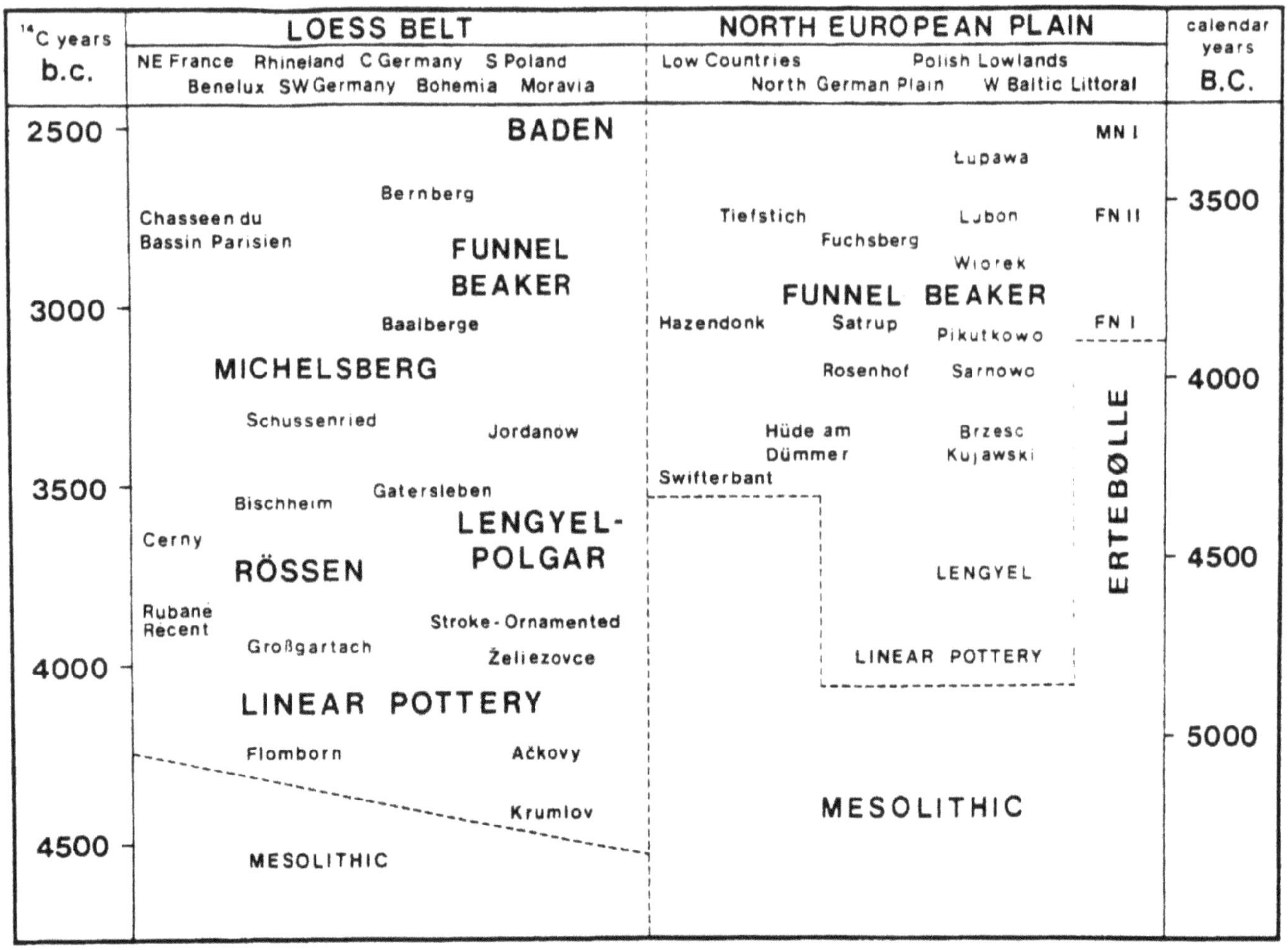

¹⁴C years b.c.	LOESS BELT		NORTH EUROPEAN PLAIN		calendar years B.C.

Figure 4.8
Chronological chart that gives some indication of the complexity of the Mesolithic/Neolithic transition in northern Europe and the development of later Neolithic groups. Formulated by Peter Bogucki (1987) the names of major archaeological cultures are in capitals while smaller cultural groups are in lower case (after Bogucki 1987:16)

by peat bog" (Bakker 1979:120). Because of wet conditions large amounts of carbonised grain as well as cereal chaff have been preserved, together with plain bowl pottery. At Hazendonk-1 large amounts of carbonised grains as well as chaff and internodes were recovered. Corrie Bakels (1998) believes the whole area to have been far too wet during the Neolithic for cereal cultivation to have been practical. She proposes that the population were importing the grain from agricultural groups in other areas. Louwe Kooijmans has also applied this interpretation to the Swifterbant groups (in Thorpe 1996:55).

The presence of cereal chaff and carbonised grain is potential evidence for the processing of grain by malting, mashing and probably fermentation as well. Small amounts of grain would have been imported and processed and it is possible to ferment a gallon of wort successfully. It is as yet unclear when these groups began to cultivate their own crops or whether they were trading with local agricultural groups. Zeiler (1991, in Thorpe) suggests exports of fish, furs and skins in return for the grain. It seems unlikely that the Swifterbant and Hazendonk groups made only flour, porridge and unleavened bread from this imported grain. The attraction of making sweet malts as well as making ale for drinking at feasts is a much more likely interpretation.

4. Neolithic groups: 4th / 3rd millennia BC

Within Europe the early Neolithic cultures, such as the Linearbandkeramik appear to have broken up around the mid 4th millennium and developed into regional cultures which are now referred to as the TRB or Funnel Beaker, Lengyal, Rossen and Michelsberg cultures (figure 4.8).

Several Scanian sites have been excavated (Jennbert 1985) where "layers containing a mixture of Ertebølle and TRB pottery" have been found. Jennbert argues for settlement and continuity between the Ertebølle and the TRB with a gradual change in lifestyle from nomadic hunter/gatherer to settled agricultural. Whether there was a rapid or a gradual change in lifestyle is currently the subject of "heated debate" (Tilley 1996:70). Christopher Tilley describes the process as "continual structuration", in which the old and the new become fused together, with a gradual alteration in the nature, scale and emphasis of activities. The gathering of wild resources would have continued as before and the cultivation of grain began on a small scale at first, intensifying over time. The new cereal products and meat from newly domesticated animals had a high prestige and symbolic value due to their exotic nature. They would have been consumed at feasts, celebrations or other events (Tilley 1996:109).

Figure 4.9
Drinking vessels, a baking plate and a bowl of the 'Funnel Beaker' or 'TRB' culture. They are not drawn to scale (after Tilley 1996: 107, source Burunhult 1982)

'Funnel Beaker' or 'TRB' culture

The 'TRB' or 'Funnel Beaker Culture' is used as a "common denominator name for a number of culturally related agricultural populations which inhabited north and Central Europe between 3600/3300 and 2150 BC" (Bakker 1979:11). It was a very widespread cultural group. Sites have been found along the northern coasts in Central Europe and also in Southern Scandinavia. As is usual in archaeology, the name of the group has come from the style of pottery that they made. This includes elaborately decorated cups and beakers, plain bowls and flat ceramics, perhaps used as baking plates or as lids for vessels (figure 4.9). Funnel beakers clearly function as drinking vessels and, given the agricultural context from which they came, they were used for milk, water or ale.

Ritual, ceremony and feasting were very important aspects of the early and late Neolithic. This is evidenced by the construction of huge communal monuments from timber and stone, for example the long mounds and dolmens (Tilley 1996:112). Evidence of votive offerings of pottery vessels, probably containing food and drink, is frequently found at entrances to the tombs. Feasts to celebrate the completion of communal constructions are also a strong possibility.

TRB pottery is found on settlement sites, at causewayed enclosures and megalithic graves or *hunebedden*, as these structures are known on the continent, over a wide area of Europe. During the 4th millennium BC causewayed enclosures and settlements begin to appear in the archaeological record, as do earthen long barrows and megaliths (Whittle 1985:186-9). As well as the construction of large, impressive tombs and monuments, the practice of agriculture became a "dominant concern" of middle and later TRB groups of the 3rd millennium BC, with increasingly elaborate ceramic forms, intensification of exchange networks and votive depositions at causewayed enclosures and at the large and elaborate megalithic tombs (Tilley 1996:119).

Artefacts that are typical of this era include flint sickle blades with sickle gloss, flint hammer stones and trapezoidal arrowheads that are found in large quantities. Saddle querns and grindstones are also found as well as flint chisels and axes of fine quality and variety in shape and style. Shaft-hole "battle axes" and double axes are more rarely found and their distribution pattern indicates the possibilities of their function as status objects. Less than sixty of these had been found up to the late 1970s (Bakker 1979:76,108).

One site that gives some indication of the complexity and the size of TRB settlements is at Hindby Moss, in southern Scandinavia. At this middle Neolithic site, 283 kilograms of pottery sherds were found representing 1049 decorated and 581 undecorated pots, as well as 22,000 pieces of flint, over 200 arrowheads, 6,000 flint cores, 142 blade knives, more than 200 flint axes, 34 ground stone axes and 7 battle axes (Tilley 1996:170). This was a large settlement, covering 20,000 square metres.

There was an explosion of pottery styles in the middle Neolithic. All point to the manufacture and consumption of a liquid product. There were drinking cups, funnel beakers and a wide variety of bowl shapes, both open and pedastalled (figures 4.10,4.11). Pottery found in funerary contexts in southern Scandinavia, such as in passage graves, survives better than that at settlement sites and appears to have taken different forms from that found at settlements, with elaborate designs and styles such as the pedastalled bowls, open bowls, ceramic spoons and brimmed beakers (Tilley 1996:255,257). All pottery is highly decorated with elaborate patterns, perhaps reflecting their use in feasting as special vessels of consumption.

For the most part the designs are abstract consisting of lines and patterns, but there is the interesting and notable exception of the so-called 'face' pots. These are found in Denmark and are attributed to the middle Neolithic III and IV (Ebbeson 1978a, Tilley 1996:273). The cups seem to have been decorated with recognizable representations of eyes, eyebrows and noses, with the handle sometimes being incorporated as the nose feature. This interpretation is open to debate. The perception of a 'face' is perhaps a subjective one. The significance is of these pots is not at all clear and just over a hundred have been found to date.

Figure 4.10
A so-called 'face pot', where the handle might represent a nose. Other similarly decorated pots have no handles. Some have 'owl-like' eyes. This example is from the Storegarden passage grave in Western Skane, southern Scandinavia (Tilley 1996: figure 6.27, no scale).

Figure 4.11
Middle Neolithic pottery vessels found in passage graves, southern Scandinavia, including pedastalled bowls, drinking cups, deep bowls and vases. This collection was in Gillhog passage grave, Skane (Tilley 1996: figure 6.7, no scale).

Pottery has been discovered in buildings that have been interpreted as being 'cult houses' or 'special buildings'. These are usually situated close by megalithic tombs. At Tustrup, Denmark, there is evidence for one of these houses measuring approximately 6 metres by 5 metres. Two dolmens and a passage grave formed a semi circular arc around the building and these were situated about 50 metres away (Tilley 1996:278). A total of 28 pots comprising 10 pedestalled bowls together with assorted beakers, bowls and clay spoons were discovered within the house, which had been destroyed by a fire, hence the survival in situ of the ritual equipment. These buildings have been interpreted as cult buildings or temples (Becker 1973) since there are no traces of human remains within them, yet they are frequently situated close by megalithic tombs, dolmens or passage graves. Ritual feasting appears to have been a probable activity within these apparently special buildings.

Emmer wheat (*Triticum dicoccum*) appears most in the archaeological record at settlement sites, the most likely place for the manufacture of all kinds of grain products. Barley and einkhorn appear in much smaller quantities. This might be because one of the products of malt and ale production is the so-called 'spent grain'. This is left after the sweet malt liquid has been washed out of the barley mash. Spent grain is an excellent animal feed for ruminant animals, such as cattle. It is not a 'waste product' and it is not likely to have been thrown away. This would be a plausible explanation for the small number of barley grains found in the archaeological record. Barley was malted, the malt sugars extracted for immediate consumption or fermentation and the spent grain would have been fed to the cattle.

Large shallow bowls and deep buckets are suitable vessels for mashing the malt and for fermenting wort into ale (see Chapter One). These styles of pottery vessels appear in abundance in the archaeological record of the TRB West culture. A wide variety of pottery styles have been found in the Dutch *Hunebedden* and at settlement sites of the West Holland TRB of the 3rd millennium BC (Bakker 1979). Middle TRB pottery styles include collared jugs, funnel beakers, lugged beakers, handled cups, buckets and a variety of bowl shapes. Such pottery styles indicate the manufacture and the consumption of a liquid product.

An interesting feature of this period in European prehistory are the deposits of "large caches of up to several kilograms of carbonised grain, often in large vessels...that are difficult to interpret...and do not appear to have been casually discarded" (Bogucki 1988:129). These apparently deliberate depositions of carbonised grain have been found at sites of the 4th millennium Lengyel culture in northern Poland, specifically, at Radziejow and Opatowice. They have been interpreted as ritual offerings and Bogucki has suggested that the stored grain "was slowly carbonised through the heat generated by its rotting in the vessel" (Bogucki 1988:134). A burning wooden building would be a more likely explanation to carbonise grain stored in pottery vessels. Carbonised grain that is discovered in a pit could indicate that kilning the malt had gone wrong, and the carbonisation occurred as a result of an accidental fire, as at the Iron Age malting and brewery site at Eberdingen-Hochdorf (figures 1.5,1.6).

In conclusion the TRB culture had the necessary material culture and pottery vessels for converting their grain into malt and ale. They had suitable buildings at settlement sites for malting, although posthole evidence on its own is often very difficult to interpret. Julian Thomas (1988) describes a "new Neolithic" on the Continent during the 4th millennium BC, with greater complexity in domestic and ritual behaviour.

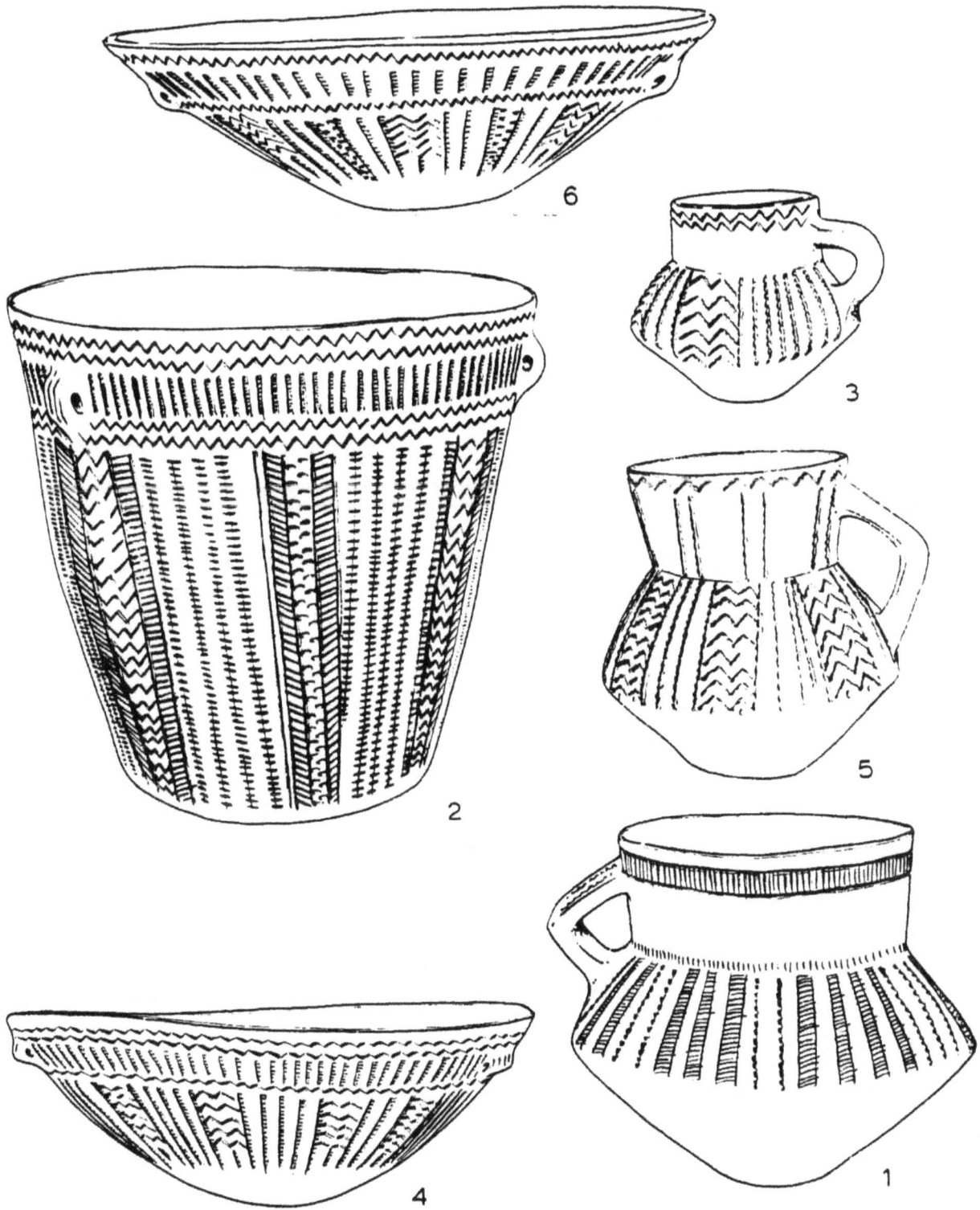

Figure 4.12
These are some of the Drouwen B TRB pottery from graves in west Holland, dated to between c3300 BC and c2100 BC. They have been categorized according to decorative style (after Bakker 1979:65). Scale is 1:3. The shape is more important than the decoration when assessing the possible function of a vessel. Here there are vessels that are suitable for mashing malt, fermenting wort and consuming the finished product. The bowl would be suitable for mashing or it could be a useful boiling/evaporation dish. The bucket would hold about 2 or 3 gallons of liquid for storage or for fermentation. It has lugs to secure a cover or lid. The cups are clearly for the consumption of a liquid product, probably mead or ale.

The development of monumentality, the building of large earthworks and central places, such as causewayed enclosures, megaliths and tombs are all features of this cultural development. The evidence for feasting and ritual in connection with these sites is abundant although the context of the feasting is not always clear. Some may have involved public festivals and celebrations, there would also have been private rituals, ceremonies and feasts and some may have included 'work party feasting', with the consumption of meat and ale for the community workforce involved in monument construction.

The evidence presented in this chapter for the manufacture of both malts and ale during the middle Neolithic in Europe is convincing. There was trade, communication and exchange with other cultures that had already learnt the techniques of cultivating cereals and processing them into a wide range of products, including ale. Suitable pottery vessels and lids were available, wide bowls for the mash and deep buckets for the ferment of a few gallons. Suitable buildings would have been available, but unfortunately, wooden buildings leave few remains, floor surfaces are often destroyed and are hard to find. The surviving archaeological evidence for the manufacture of malt and ale in the Neolithic is much better on the Orkney Islands, Scotland, where the buildings were constructed from stone. Unstan pottery is bowl-shaped, similar to the Drouwen bowls above and Grooved Ware is deep, large and bucket-style, also as above.

The evidence for the malting, mashing and fermentation at settlement sites of the Orcadian Neolithic is the subject of the next chapter.

CHAPTER FIVE
The Stone Buildings of the Orcadian Neolithic

1. Neolithic settlements of Orkney: 4th/3rd millennia BC

The most complete and the best preserved domestic buildings dating to the Neolithic in the whole of northern Europe are to be found on the Orkney Islands, at Knap of Howar, Barnhouse and Skara Brae. Here the houses were constructed of stone rather than timber. At all three sites barley and wheat were being cultivated and processed during the 4th millennium BC.

Knap of Howar, on the remote northern island of Papa Westray is the oldest standing house in northern Europe. Radiocarbon dating indicates that the house has a chronological range from c3800 BC until c2800 BC, although the actual length of habitation may have been for only around 500 years (Ritchie 1983:57). Other settlement sites on Orkney are not as well preserved. Barnhouse (Richards 1992) is situated close by the Stones of Stenness and it has been dated to the mid 4th millennium BC. Only the wall footings remain because of intensive ploughing of the land.

Skara Brae is situated at Skaill Bay on the west coast of mainland Orkney. The village is very well preserved because it has been buried for millennia beneath sand dunes. The stone buildings and artefacts were first revealed in the 19th century, after a particularly severe storm. The village was initially investigated by Flinders Petrie (1867). Gordon Childe continued excavating and recording the village between 1927 and 1929 (Childe 1930). Both archaeologists thought it was a 'Pictish' village but radiocarbon dates obtained in the 1970s during David Clarke's excavations revealed the true antiquity of the site (Clarke 1976). The village was Neolithic and had been inhabited between c3000 BC and c2000BC.

The site of Rinyo, on the island of Rousay at Bigland Round, is now completely hidden under the grass. Unless new excavations are undertaken, all that remains for study are the ground plans that were drawn up during Childe's investigations and the photographs taken during the excavations (Childe and Grant 1938, 1946).

Other Orcadian Neolithic settlements that have been excavated but have not yet been fully published include the extensive site at Links of Noltland, Westray, and other fast eroding coastal sites at Pool, Tofts Ness and Stove Bay on Sanday (figure 5.1). Two early Neolithic settlement sites on mainland Orkney have recently been excavated (Richards forthcoming). These are situated close by the tomb on Cuween Hill and at Stonehall, a few miles west of Kirkwall, by the Stromness road.

The building remains at Knap of Howar and Skara Brae are unique. These and other Orcadian Neolithic settlements can provide valuable clues to the activities of Neolithic cultures throughout the British Isles and Europe.

Figure 5.1
Map of the Orkney Islands, showing the sites discussed in the text.

41

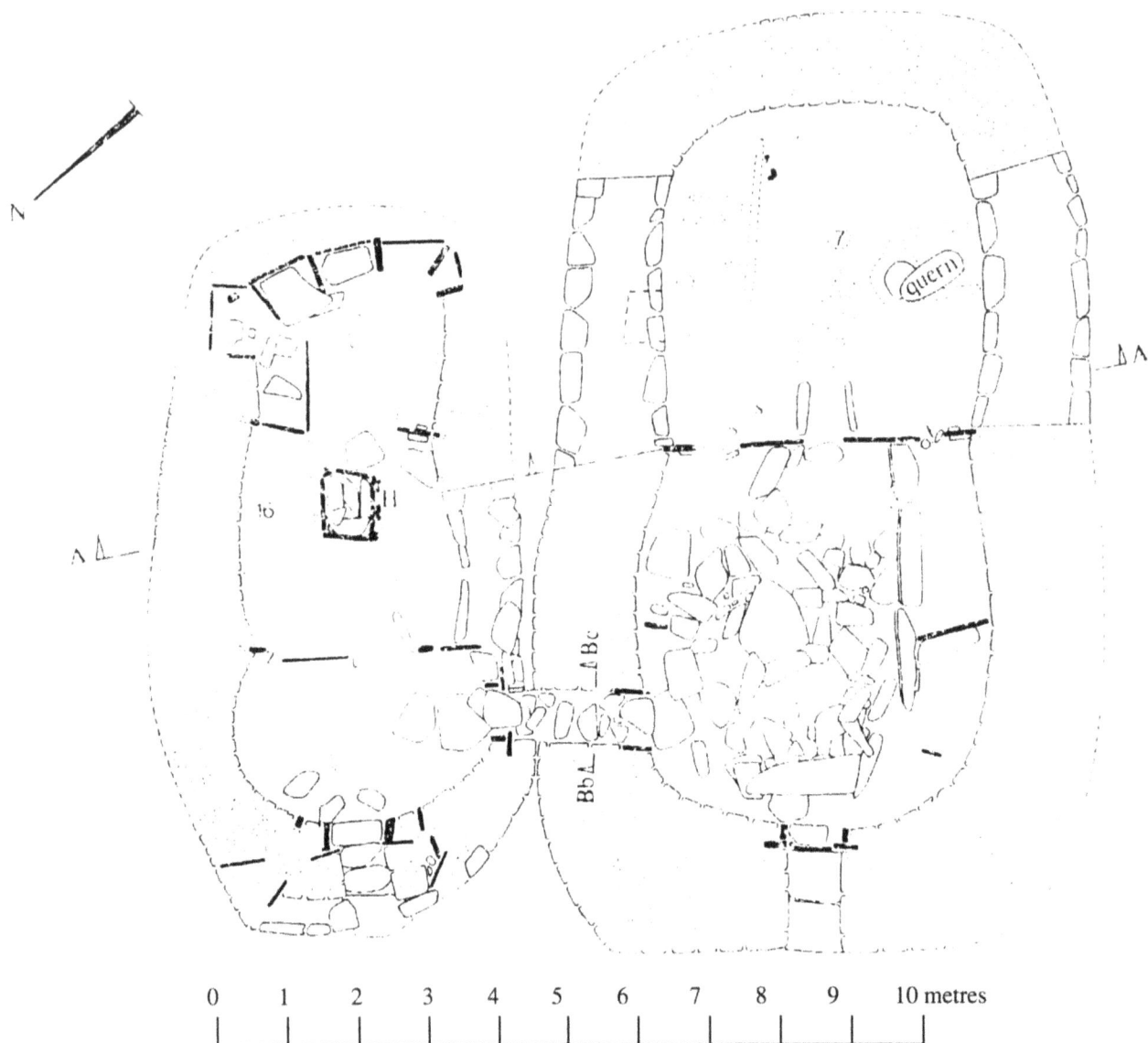

Figure 5.2
Ground plan of the buildings at Knap of Howar, Papa Westray, Orkney (Ritchie 1985:47, figure 3).

Knap of Howar, Papa Westray

According to radiocarbon dates the oldest settlement so far discovered on the Orkney Islands is at Knap of Howar on Papa Westray, one of the most northerly islands of the Orkney group. Here, a dwelling house and workshop, both stone built and connected by a narrow passage, were discovered and initially investigated in the 1930s (Traill & Kirkness 1936). Detailed excavations have been carried out by Anna Ritchie in the early 1970s when radiocarbon dates from between c3800 and 2800 BC were obtained (Ritchie 1983).

House 1, the larger of the two connected buildings, is often referred to as being the dwelling house while House 2, the smaller building, has been interpreted as a workshop and storage area (figure 5.2). The reality probably was that both buildings had several functions, according to the time of year and the necessary activities undertaken. According to the excavated evidence of a wide variety of fish bones, cattle, sheep and pig bones and carbonised grain found in the extensive midden that surrounded the buildings, the people who lived on Papa Westray 5000 years ago were adept at

fishing, both onshore and offshore, and practised animal husbandry. They cultivated and processed grain as well (Ritchie 1985:22). The pottery assemblage consisted mostly of wide, shallow, highly decorated bowls known as Unstan Ware (figure 5.4). Sherds from deep, flat-based storage vessels were also found (figure 5.3). Unstan bowls would be suitable for mashing the malt and the deep vessels would be ideal for the fermentation of wort, if covered. Quantities of grain processed would probably have been small.

Having visited and studied the buildings at Knap of Howar, it seems that the smaller building (referred to as House 2 in Ritchie 1983) would have functioned well for food preparation, cooking, storage, eating and sleeping (figures 5.5-5.8). The building has been separated into three distinct areas using upright stone slabs set into the ground. There may have been screens made from organic materials as well.

Several cupboards and storage places were built into the walls, particularly in the innermost, rear part of the building (figure 5.7,5.8). In this rear section there are also suitable places to make safe and undisturbed sleeping places for babies, infants and small children. There is enough floor space for two adults to lie down in this area as well.

The hearth is stone kerbed and centrally located, a practical area for food preparation. A quern was located in this section of the building. In the front entrance area there is a narrow passageway that connects the two buildings together. Access through this passage was controlled by a door within this, the smaller building of the two. On excavation, the connecting passage was found to have been deliberately blocked with stones (Ritchie 1985:27). Reasons for this are as yet unclear.

The larger building (referred to as House 1 in Ritchie 1985) is divided into two separate areas, an open hearth being located in the rear section, together with a very large quern. This was apparently still situated in the place where it had once been regularly used (figure 5.12). I think that this larger building is suitable to have functioned as a barn for animals and a building for grain storage and preparation.

The front area was divided from the back with upright stone slabs and has a very low stone construction, described as a 'bench' or 'platform'. It is too low to be comfortable to sit on and it is reminiscent of the floor construction of cattle stalls at the animal barn at Corrigall Farm Museum, Harray, Orkney. There was a paved stone floor (now covered by gravel) in the front area. Animals could have been kept and tended to here and the harvested grain could have been stored, malted and further processed at the rear of the building. The large upright slabs and postholes dividing the building would have been an effective barrier, keeping the animals away from the grain storage and preparation area.

There are two entrances or doorways into House 1 (figure 5.10). One leads in from the outside and the other is an entrance from the passage connected to the smaller building. The main doorway into House 1 is quite significantly larger than the doorway leading in from the passage. Could this have been a deliberate design feature? It would mean that large animals could enter the barn through the main door but could not gain access into the habitation area via the smaller doorway and narrow connecting passageway.

My interpretation of the buildings that make up the early Neolithic farmstead at Knap of Howar is that House 1 may have been the barn, being divided into a grain storage and processing area at the rear and an animal shelter at the front (figures 5.11, 5.12). House 2 is the living, cooking, sleeping and general working and storage space.

The design of this building is unlike those at Skara Brae and Rinyo. This has contributed to a belief that there were two different cultural groups on Orkney during the Neolithic. There were the people who used Unstan Ware pottery, like that predominantly found at Knap of Howar, and those who used Grooved Ware, as at Skara Brae and Barnhouse. This division of the Orcadian Neolithic into two separate cultures is questionable and has been debated (see Sharples & Sheridan 1992). It is not viable to divide the Orcadian Neolithic in this way. Both pottery styles are represented at Knap of Howar. Sherds of large flat-based vessels were found there and, though not strictly Grooved Ware, they are similar in size, shape and probably also

function.

Both Unstan and Grooved Ware have been found at Neolithic settlement sites, Pool and Stove Bay, on the island of Sanday (Bond 1995). The potential function of the two pottery styles is different. Wide, shallow bowls would be suitable for a variety of activities including the boiling or evaporation of liquids and mashing the malt. Deep, large bucket shaped vessels would be useful for storage and, with the addition of a lid, for fermentation of wort.

At Pool, Sanday, the remains of 14 houses have been found together with sherds representing almost 2000 pottery vessels. Unstan Ware predominated in the lower levels while Grooved ware predominated in later levels. Perhaps the emphasis on the different grain processing activities changed, with more malt/sugar extraction in the early Neolithic and a greater amount of fermentation in the later Neolithic as the amounts of barley cultivated by the inhabitants increased in quantity and their techniques improved. There is clearly much more research needed into the potential functions of pottery vessels in the Neolithic.

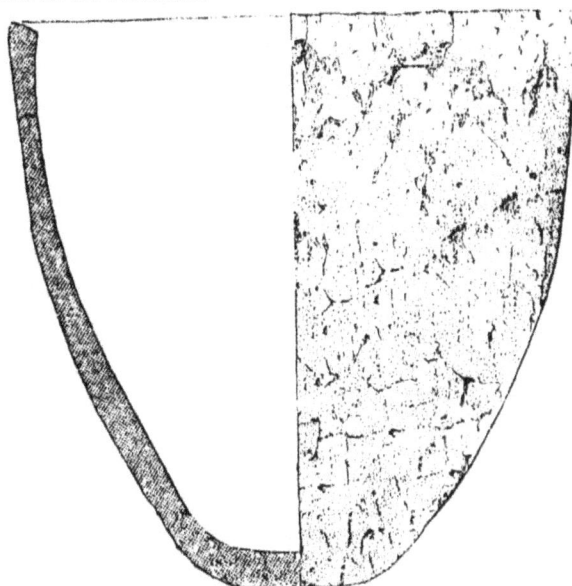

Figure 5.3
Large, deep pottery vessel from Knap of Howar (after Ashmore 119:46, figure 21).
Rim diameter is approximately one foot.

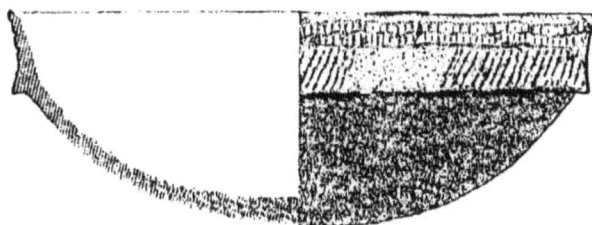

Figure 5.4
Wide, shallow bowl from Knap of Howar (after Ashmore 1996:46, figure 21).
Rim diameter is approximately one foot.

The following photographs of the buildings at Knap of Howar, Papa Westray, illustrate aspects of the interpretations discussed above.

Figure 5.5
The smaller building (House 2) at Knap of Howar, Papa Westray, Orkney, showing the entrance, upright slabs dividing the front and middle sections of the building, the central stone kerbed hearth and main food preparation area with a quern stone on the left of the picture.

Figure 5.6
View of House 2 from beside the doorway, showing the middle and rear sections of the building. The rear of the house would be the most sensible area to use as a sleeping or resting area, being the place where one would be least disturbed.

Figure 5.7
This picture shows the areas constructed at the rear (left) of House 2 that may have been used for sleeping or resting. There are niches in the back wall, perhaps for babies and small children, as well as two spaces beside the upright dividing slabs that are large enough for adults to lie down in.

Figure 5.8
Storage places rather than sleeping areas have been incorporated in the wall construction at the rear (right) of House 2.

Figure 5.9
The passage leading from House 2 into House 1, to the right of the main entrance to House 2. The passage is paved with flagstones and there may have been a door on this side. The height of the doorway is about one metre.

Figure 5.10
This picture shows the different sizes of doorways in House 1 at Knap of Howar, Papa Westray. The smaller door (right) leads to the passage into House 2 and is about one metre high. The main entrance to House 1 is significantly larger, perhaps to allow access to large animals. The floor was originally paved with flagstones. This section of building appears to be more suitable for keeping animals than for people to live in.

Figure 5.11
The low stone construction on the floor in House 1, facing the passage between the two buildings. It appears to be the footings of an animal stall with an upright slab dividing it into two unequal areas. It may have been used as an area for milking or feeding the animals.

Figure 5.12
The grain processing area at the rear of House 1, Knap of Howar. There is a very large quern stone and stone rubber, probably still in situ. There are footings for postholes and upright slabs that, with the addition of a central wooden gate, would be effective in keeping the animals in the front section, well away from the grain processing area. House 1 is much larger than House 2.

Barnhouse, mainland Orkney

Excavations at Barnhouse Farm, mainland Orkney undertaken by Colin Richards and a team from Glasgow University have revealed a large, multi-phase Grooved Ware settlement (Richards 1992). Here, only half a mile from the Stones of Stenness, there was a village of about 15 houses. The village is contemporary with the stone circle, that is, the mid 4th millennium BC. Many large Grooved Ware vessels, some still intact, were discovered in situ where the villagers had used them. The context of this Grooved Ware pottery may allow the function of the pottery to be assessed. Unfortunately the buildings at Barnhouse have not been preserved as well as those at Knap of Howar and Skara Brae because the site has been ploughed with machinery.

Building 8

This is the largest Neolithic building yet excavated on mainland Orkney. It is also the most unusual. It has been interpreted as a temple by the excavator, Colin Richards (1992). Other interpretations agree that this building probably had a special function that was more to do with ritual activity than habitation and everyday domestic life (Barclay, Darvill, Topping, 1996). It is a roughly square building and has been described as "centrally positioned on a sub-circular clay platform enclosed by a stone wall" (Barclay 1996: 68). The walls were substantially constructed, the outer wall being 1.5 metres thick and the inner wall being 3 metres thick. The floor area of the 'inner sanctum' is roughly 60 square metres.

It may be possible to interpret the area of clay flooring between inner and outer walls as being suitable for a malting floor: it fulfils the necessary criteria but, at present, such a suggestion can only be a supposition pending further scientific investigations of grain found there. There are two hearths within the outer area, both by the drain. This area might be useful for processing activities such as mashing, an activity that requires access to fire, water and drainage.

Within Building 8 the footings for a stone 'dresser' were found, even though the surface features of the whole site have been badly damaged by ploughing over the years. Two hearths were also located during excavations, one centrally placed and the other apparently placed in the entrance to the building. Building 8 has been so reconstructed that one has to actually leap or step over the hearth in order to enter the building. Richards argues that such unusual aspects indicate that the building was, in all probability, constructed with ritual activity in mind and he believes that these rituals involved the use of fire and water (Richards 1992). These two elements are of great significance in the transformation of grain into malt, malt sugars and ale. Could this have been one of the ritual activities undertaken in this building?

Excavations at Barnhouse have uncovered several drains. In Building 8 the drain runs under the "dresser" and under both internal and external walls. It could have served to channel waste liquids outside the building. Drains have been found at Skara Brae, Rinyo and at other recently discovered and excavated Neolithic sites, for example, Cuween Hill and at Stonehall (Richards forthcoming). Clearly the Orcadian Neolithic people planned, designed and constructed their settlements, houses and other buildings with integral drainage systems. Why did they do this and what was the function of the drainage system so carefully planned and built at so many

settlements? Practical experiments have shown that malting, mashing and fermentation require access to and the use of a lot of water. Waste liquids, such as the washings from pots, have to be disposed of. The necessity for washing out and cleaning the used sticky vessels and equipment from malt sugar manufacture is a major and essential part of the whole process of brewing. If the vessels and containers are not kept scrupulously clean then subsequent attempts to make ale in that vessel will fail and the ale will be sour and undrinkable. Cleanliness of equipment is crucial for successful brewing.

The presence of drains at so many of these Orcadian Neolithic settlement sites is a necessary feature for brewing to have taken place there. For any brewer, even today, the drains are essential and necessary. They are an important and a significant feature as a visit to any brewery, large or small, will demonstrate.

Building 2

Here, evidence was found for the removal of barley husks, particularly around the eastern hearth (Jones 2000). It is the only building that was in use throughout the life of the settlement. Other houses were rebuilt, altered or abandoned over the years (Richards 1992). The constant use of Building 2 might reflect its special function and importance as the grain processing centre for the whole village. Upright slabs were used to separate different working or living areas. There were two large stone-kerbed hearths.

Ritual activity within Building 2 is hinted at with the female burial in a cist, positioned almost in the centre of the building by the outer hearth and so placed that anyone who entered could not avoid stepping on and over it. Patrick Ashmore has suggested a 'cult use' of the building, in terms of the high quality stone tool, mace head and carved stone ball production that took place there (Ashmore 1996:50). A substantial stone-built drain runs out of the building, from the north corner. Most of the buildings at Barnhouse had stone drains that fed into Loch Harray, a freshwater loch (Ashmore 1996:49). Drains are essential for wet grain processing.

The village was a well-organised community. There was a central outdoor workplace for making flint and stone tools, for the manufacture of ceramics and for skin and hide preparation (figure 5.13). These are essential activities that could be easily done out of doors, weather conditions permitting. Grain processing, however, would have to have taken place within a building, to afford the malt suitable protection from the elements and from animals and birds.

With its location in the heart of the ritual centre of Orkney, close by the two stone circles and within sight of the chambered tomb of Maes Howe, Barnhouse as a settlement appears to encompass both ritual and domestic aspects of Neolithic life. Suitable equipment and conditions for malting, mashing and brewing can certainly be found there. There were large Grooved Ware vessels, evidence for grain processing activity in Building 2 and a complex and apparently planned drainage system to serve the whole settlement. All these are elements related to and necessary for the ritual of brewing.

The Grooved Ware pottery has been analysed using the scientific analytical technique of Gas Chromatography and Mass Spectrometry (Jones 1999, 2002).

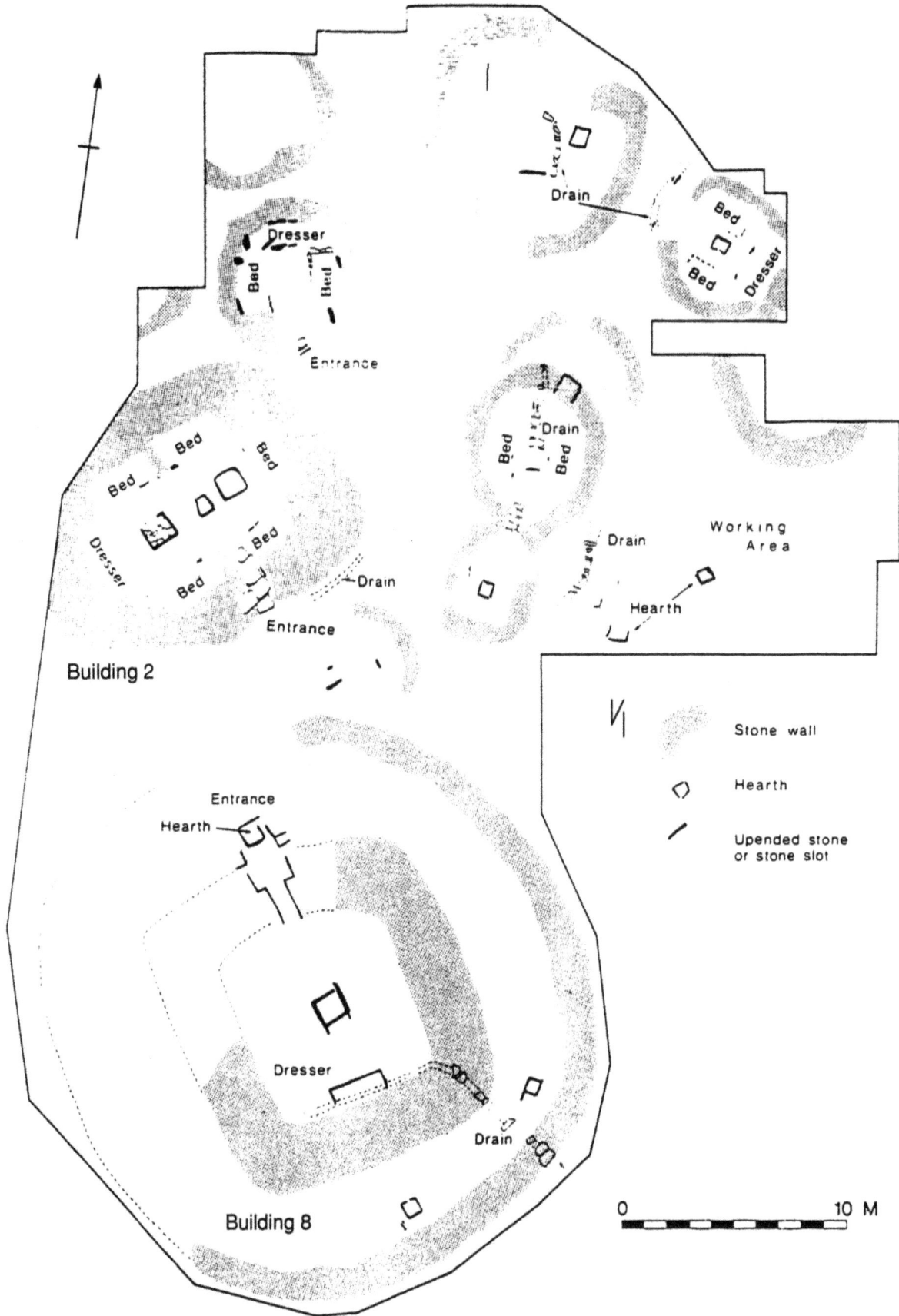

Figure 5.13
Ground plan of the settlement at Barnhouse, Orkney, showing the central working area, the drains from Buildings 8, 2 and other buildings (Bewley 1994:58, courtesy of Richards). The drain leading from the north corner of Building 2 is not shown on this plan although the drain from the entrance is shown.

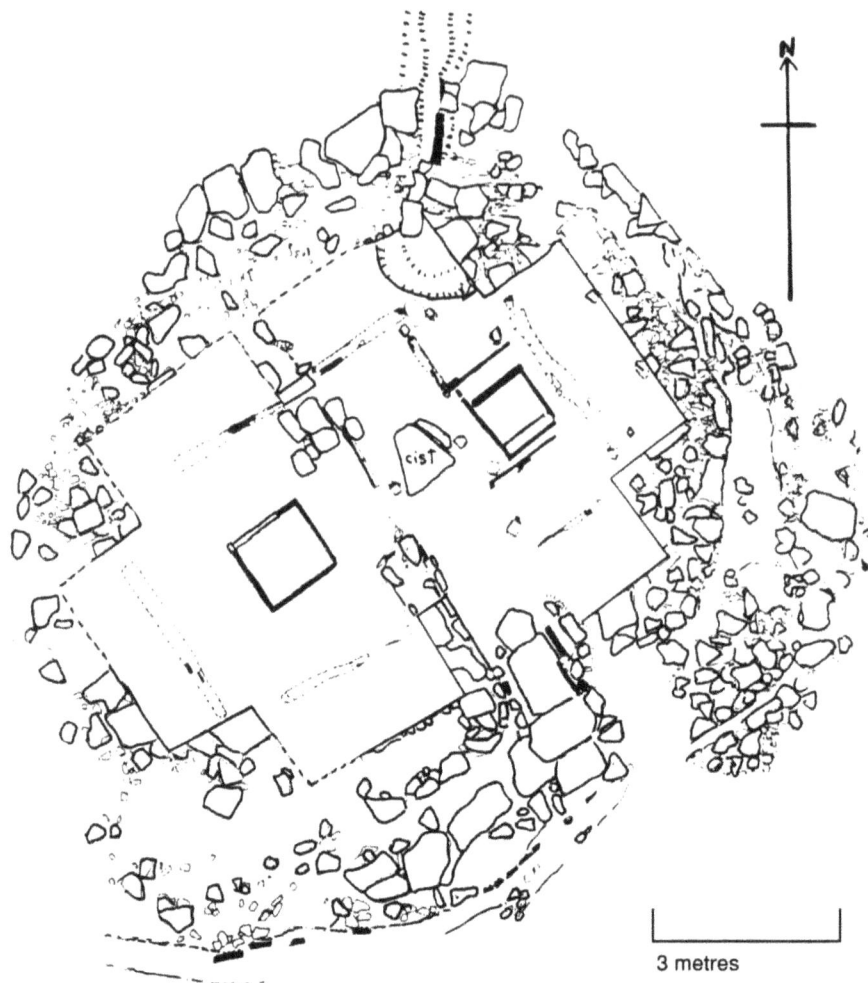

Figure 5.14
Building 2, Barnhouse, Orkney, showing the drains from the northwest corner of the building and by the entrance. Drains are an essential requirement for brewing. This building was probably used for wet grain processing activity and also for the manufacture of special stone objects. A female skeleton was found within the cist. (after Bewley 1994:58).

Using this method organics and lipids can be extracted from the fabric of ancient ceramic vessels and different classes of food can be identified by their different chromatographic signatures. Initial analysis has revealed that the Grooved Ware vessels at Barnhouse were associated with barley, 'unidentified sugars', cattle milk and cattle meat. There was also evidence that the sugars may have been associated or mixed with milk in some way (Jones pers comm).

The presence of sugars could be an indication of mashing, that is, the extraction of malt sugars and malt liquid from the barley grain. However, it has to be also considered that these sugars may perhaps be derived from milk products, although this is unlikely. The possibility of mixing malt sugars with milk is an interesting proposition. Mixing the malt extract with milk would have made a nutritious drink, very like malted milk drinks that are popular today, such as Horlicks. This would have been a welcome drink in the cold of an Orkney winter and probably appreciated more than ale. It would have made an excellent and nutritious drink for everyone, including children and infants, providing them with

essential B-vitamins from the malt and calcium from the milk. The ritual of grain processing at Barnhouse involved not only brewing but also the malting and mashing of grain into malt sugars, thus producing a nutritious and valuable dietary supplement for all, adults and children alike.

At Barnhouse, Dr Jones' research revealed that the medium sized Grooved Ware vessels had contained either milk or meat. If vessels had been sealed prior to use with animal fats or butter then these results might perhaps be indicative of this treatment. The strongest evidence for barley apparently came from the smallest and the largest vessels. These are interpreted as being the serving and storage vessels respectively, particularly since the largest vessels were in static locations (Jones, pers comm).

There is a possibility that the large vessels at Barnhouse were used for the fermentation of barley wort. The weight of a large Grooved Ware vessel filled with fermenting liquid would make it impossible to move, hence their static location. Liquids can be transferred from one vessel to another by siphoning, using reeds or by using smaller vessels.

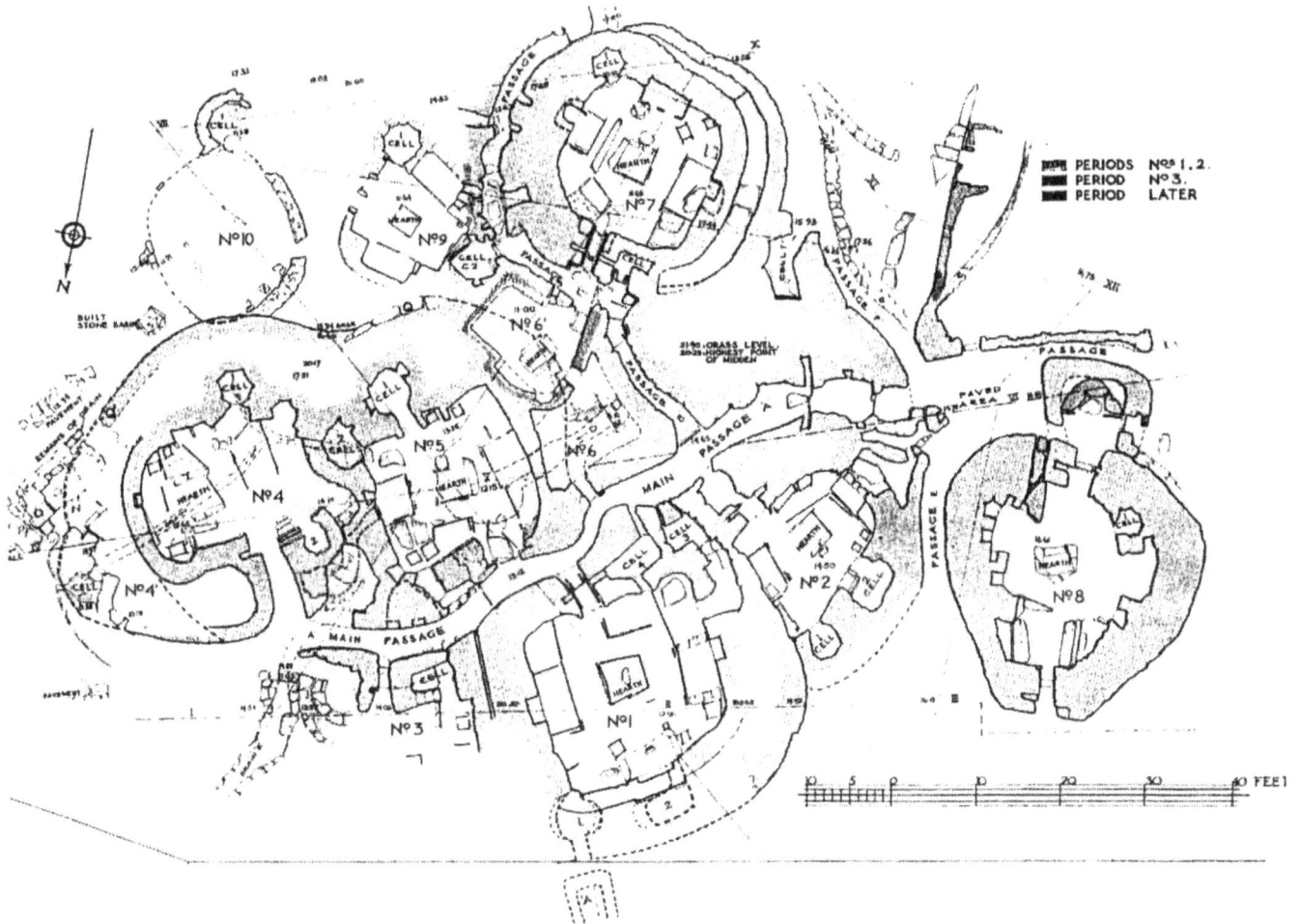

Figure 5.15
Ground plan of Skara Brae, Skaill Bay, Orkney (Childe 1931). The drains are shown as dotted lines beneath and outside Hut 4 and outside Huts 3 and 1. During Childe's excavations green sludge was discovered in the drains and within Pit P and Pen D in Hut 7 (see figure 5.19). There was a large amount of it beside Wall Q, in the area outside Huts 4, 5, 9 and 10. Two female skeletons were found buried in a cist beneath Hut 7.

Storage of the barley grain at other stages of processing as well as the storage of other foodstuffs could also be functions of these large static Grooved Ware vessels. According to Dr Jones' research, the best evidence for barley came from the complete medium sized Grooved Ware pot that was discovered sunk in the ground up to its rim in Building 8. Could this have been a 'must pot', that is, a pot maintaining a yeast culture? Storage vessels that are set into the ground have a steady temperature and would be perfect for the maintenance of a yeast culture. All the elements necessary for malting and brewing are to be found at Barnhouse. There were floors suitable for malting, stone built drains, suitable pottery vessels and suitable buildings, as well as the evidence for the 'removal of husks' in Building 2 (Jones 1999).

Skara Brae, mainland Orkney

A violent storm in the 1850s revealed middens at Skaill Bay that were 15 to 16 feet deep and contained artefacts of stone and bone. The discoverers, Mr Watt and Mr Farrer, began initial investigations and discovered that the stone buildings had also been preserved beneath the sand dunes. George Petrie began excavations in the 1860s and published the first plans of the buildings, now known as Huts 1 and 3 (Petrie 1867). He discovered a drain leading out to sea from a rectangular cell (A) attached to the seaward side of Hut 1 (ibid:205). This part of Hut 1 has since been destroyed by the sea but it is included in Childe's ground plan (figure 5.15).

Artefacts found at Skara Brae included a quartz celt, whale rib bones, which may have been used for roof supports, carved stone balls, several circular stone pot lids and sherds from very large pottery vessels. Petrie was impressed with the "considerable constructive skill" of the builders of this ancient settlement (1867:217).

Properly recorded excavations did not begin until the mid 1920s, by which time storms had washed away much of the remains of Hut 3 and also the drain and the rectangular cell outside Hut 1, discovered by Petrie. Vere Gordon Childe was called in to supervise the consolidation of the buildings and to investigate and excavate this most unusual site. During 1925 and 1926 the sea wall was constructed to protect the site from further damage and destruction.

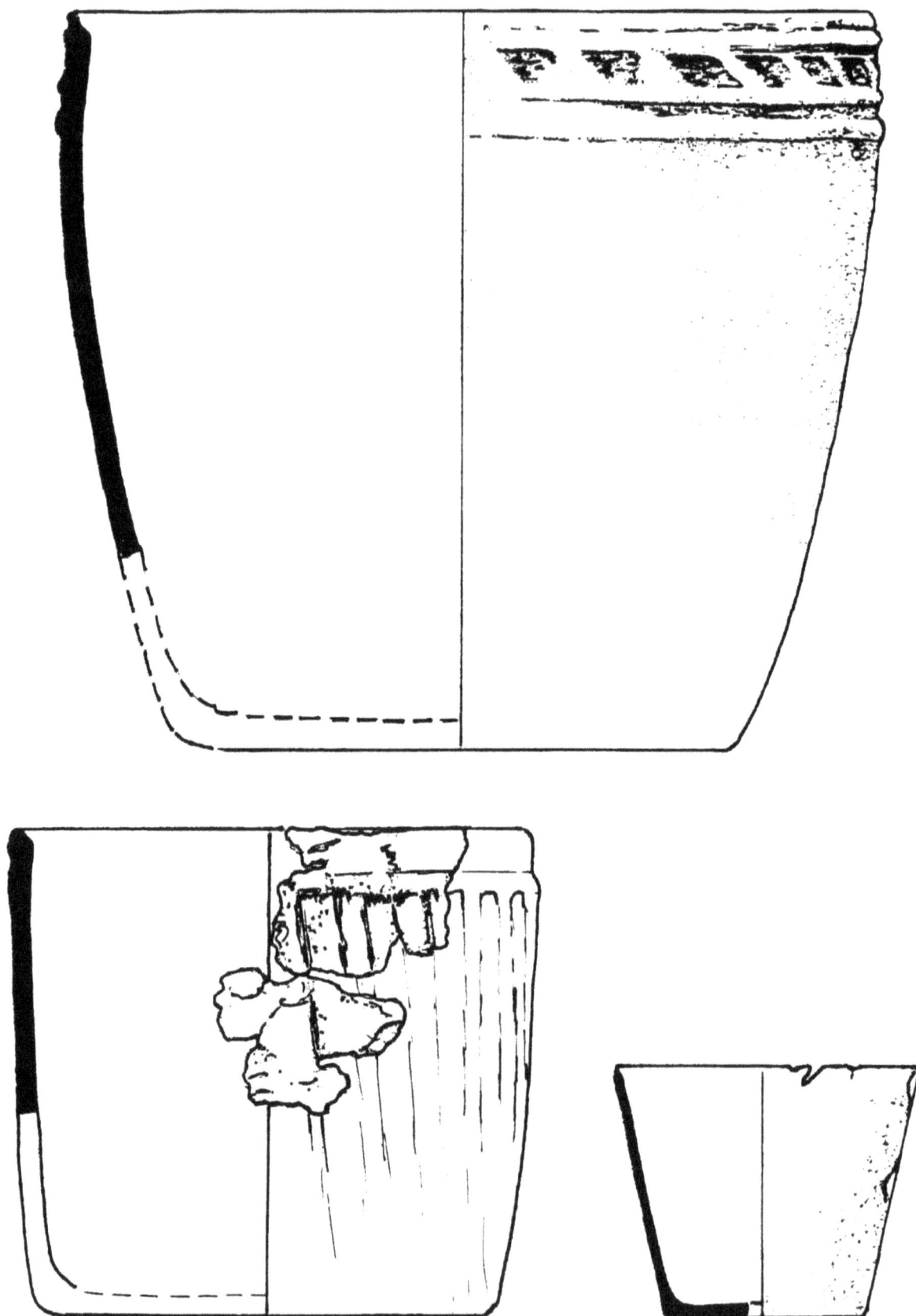

Figure 5.16
Grooved ware pottery vessels from Skara Brae discovered during Clarke's excavations in the 1970s (after Clarke 1976).
Scale 1:5

The largest vessel could have held up to about 30 gallons of liquid. Circular stone pot lids were found at Skara Brae and
would provide suitable anaerobic conditions for fermentation of barley wort. See Chapter One for details.

Excavations at Skara Brae began in 1927, when Mr. J. Wilson Paterson supervised work, and continued through 1928 and 1929 under the supervision of Gordon Childe. Detailed ground plans and sections of Skara Brae were drawn by Mr. J. Houston of the Office of Works, Edinburgh. These are invaluable in understanding the stratification and the relationships between the different buildings. The ground plans and, in particular, the many sections drawn of Skara Brae merit close study, as do the detailed descriptions (Childe & Paterson 1929, Childe 1930). Childe and Paterson's work was much more than a simple conservation exercise.

Attempts were made to assess "stratigraphical variation" (Childe 1929:239), to plot finds and record accurately the settlement that they were investigating. At times it was a confusion of walls and midden material and it was all very difficult to interpret. Several levels of occupation were revealed by initial investigations under the supervision of Gordon Childe and Paterson and more recently by David Clarke's excavations in the 1970s (Childe 1931; Clarke 1976b). Full details of Clarke's excavations are not yet available, although summary accounts have been published (1976a,b). Clarke's excavations provided organic material for radiocarbon dating. Carbonised grain was found in the middens. Dates for the first period of occupation range from 3380 - 2585 cal BC, and for the second period from 2585 - 1975 cal BC (MacSween 1992: 268).

Childe described his impressions of "hasty abandonment" and the evidence of later different and interspersed use of the settlement. With no apparent trace of hostile activity, it is most likely that the settlement was overwhelmed periodically by violent storms (Childe 1931:61). Childe interpreted four re-occupation levels in Hut 7, with layers of sand and deer bones in the upper levels indicating the continued use of the site after the terrible storm which appears to have killed the cattle and destroyed the crops of the early inhabitants. Hut 7 was filled with sand and, as Childe dug down, he discovered a square stone hearth at a depth of 5 feet from the roof level and several feet above the floor. This had been constructed upon windblown sand. Red deer bones were found both above and below this hearth (Childe 1929:249).

The early inhabitants grew barley and wheat, kept domestic cattle and pigs and were skilled seafarers, having arrived by sea to the Orkney Islands at some time in the early 4th millennium BC (Ritchie 1985:21), presumably bringing with them cattle, grain and the knowledge and rituals of grain processing techniques. Their skills in crafts are indicated by the fine, elaborate tools and pottery that were made there. Excellent engineering and planning skills are demonstrated in the construction of the buildings and the drainage system. Ten buildings have been uncovered, with evidence of rebuilding and refurbishment to all except Hut 7, which stands on the natural clay (Childe 1929:251) and which was in use throughout the life of the settlement. The furnishings in all huts indicate organised and structured living. Most huts have stone 'dressers' and 'beds' and some have stone-lined and clay sealed boxes built into the floor. Not all the hut interiors are identical. The internal furnishings of the huts are variable in design and therefore the huts varied in function and usage also (Barclay 1996:67).

Some of the pottery vessels at Skara Brae were very large indeed, with rim diameters up to 2 feet and measuring up to 2 feet deep (figure 5.16). Circular pot lids were fashioned from stone to fit these pottery vessels, some of which could hold up to 30 gallons of liquid. Such large pots were made for a purpose by skilful potters and they must have been clamp fired outside the buildings. The pots are too large and too wide to have been moved in through the doors of the huts and it is most likely that they were placed into the huts using access through the roof and were then left in situ. The most likely function for these huge pottery vessels would be as storage vessels or for the fermentation of barley wort.

The new Visitor Centre at Skara Brae (constructed 1999) lists 'beer' as being amongst the beverages likely to have been consumed by the early inhabitants at Skara Brae but no suggestions are presented as to the techniques and methods used to manufacture it. Clarke and Maguire (1989:22) have also suggested that 'beers' may have been made, although they seem to be hinting that it would not have been 'proper beer' but an inferior product. It is true that it would not have been a bright, fizzy, carbonated beer as we know it today. Instead, it would have been fairly clear, flat, uncarbonated ale that was flavoured and preserved with Meadowsweet or other local herbs and made in a similar way as that described in Chapter One.

This ale was not an inferior product. It is a traditional ale, as has been made and consumed over millennia and one that can still be made today with the right knowledge and skill, suitable ingredients and basic equipment. Henbane was apparently another herb that was used by the inhabitants of Skara Brae. Perhaps it was used for toothache, as is suggested at the Visitor Centre. Perhaps it was a ritual additive to ale, as discussed earlier with reference to the Balfarg residues.

Hut 8, Skara Brae

The excavated contents included numerous chips and cores of chert, masses of clay, piles of heat cracked stones, pots and lids, stones and pounders and stone knives (Childe & Paterson 1929). Large Grooved Ware vessels stood in the Hut and in the porch, which was a later addition to the building. An internal wall, to the right of the door when entering, was also added at a later date to the construction of the original building (Childe & Paterson 1929:174). The building has no 'bed' compartments and no 'dresser' but it does have numerous shelves, ambries and footings of a kiln.

At the north end of the hut, there are the remains of a structure described as "... a built wall projects at right angles to the hut's east wall, but is not properly bonded in therewith. After 2 feet it turns north again, but its line is continued by a wall of thin slabs set on edge. Beyond this partition lies a complex structure, which, as Mr Paterson cleverly points out, bears a close resemblance to a kiln. On the other three sides this annex is bounded by big flagstones set on edge, the corners being rounded off with courses of dry masonry. There is a gap between the slabs forming the rear (northern) wall and a similar gap flanked by another pair of slabs set on edge in the outer wall. Between the two pairs the gap is traversed by a low wall supporting a lintel slab. Mr Paterson interprets this as a kiln flue ..." (Childe and Paterson 1929:176). This description is invaluable as an eyewitness account. Paterson's interpretation of it as a kiln is

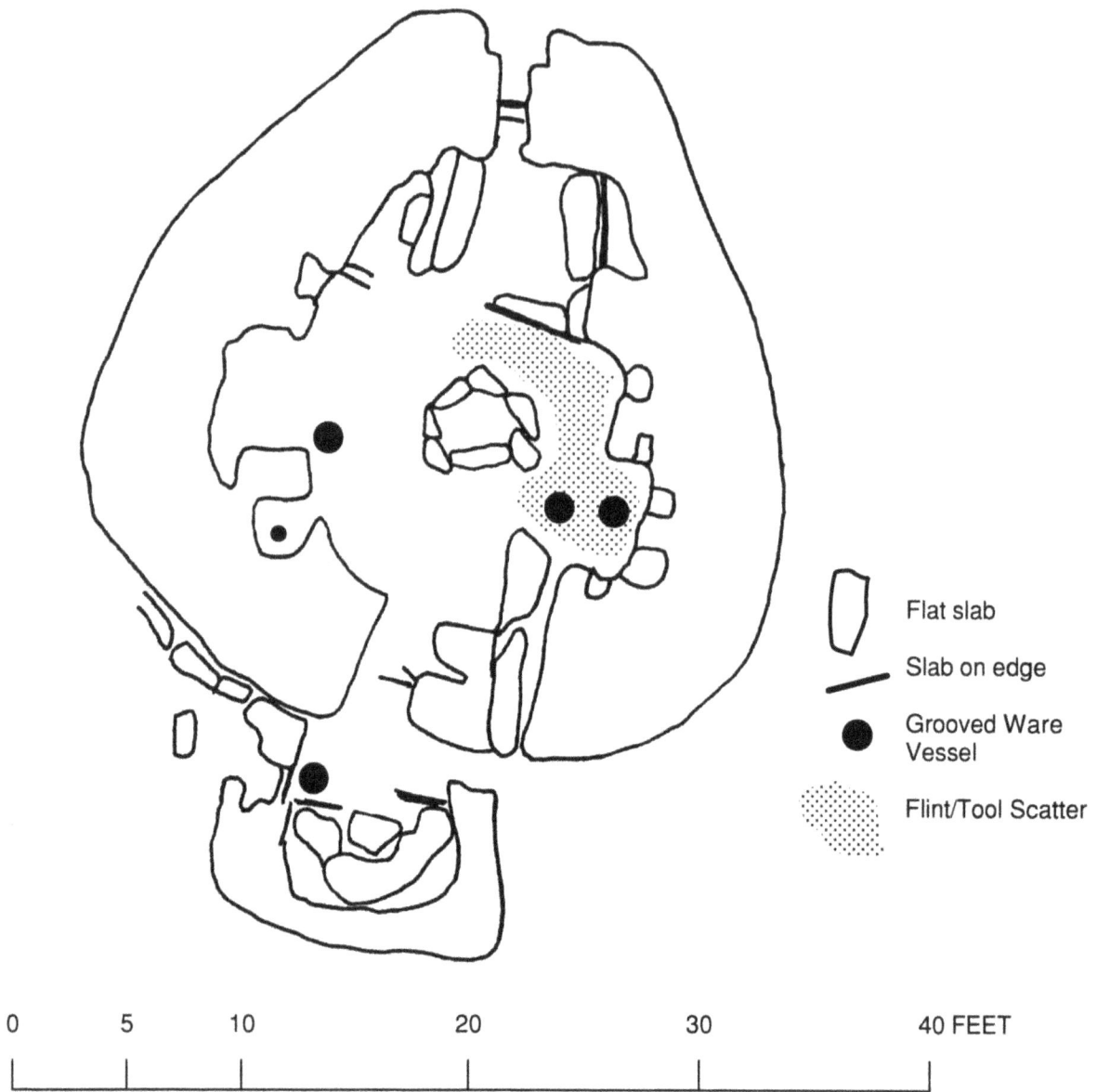

Flat slab

Slab on edge

Grooved Ware Vessel

Flint/Tool Scatter

0   5   10   20   30   40 FEET

Figure 5.17
Ground plan of Hut 8, Skara Brae.

Five large Grooved Ware vessels were found within this building. One was situated in the porch, one was in an alcove and the other three vessels were in the main part of the building. Childe reported that there was an area of flint and chert to the right of the hearth. This area was probably used for tool manufacture and repair. The porch appears to have been added at a later date to the construction of the main building. The opposing doorway and kiln flue would have provided a through draft for winnowing the harvested grain. Stone slabs on the floor and other surviving parts of the kiln can be seen by the north-facing aperture in the wall. This aperture may have been specially designed and built to 'catch' the wind. Many heat-cracked stones were found by the kiln (after Childe 1931, traced from the groundplan and enlarged).

supported by the discovery of oven or kiln bases at the Neolithic settlement of Rinyo, on Rousay, (figure 5.27). The necessary materials, equipment and conditions for malting grain and drying malt can all be found within Hut 8 at Skara Brae. The floor is potentially suitable for malting. There were stone pounders for cracking the malt, large vessels with lids for storage and a kiln in which to dry the malt. In many respects Hut 8 has similarities to the barn at the Corrigall Farm Museum (figure 1.3) where the Bere Barley was malted for the brewing experiments that were undertaken as a part of this thesis. At Corrigall, the winnowing area is located between two opposing doorways, using the wind that is channeled through to separate the grain from the chaff by throwing it into the air. The grain, being heavier, drops to the floor while the lighter chaff is blown away, out of the barn (Flett, pers comm). The person performing the task of winnowing can do so in a warm, dry place and good use is made of the wind in wet weather. It is practical and expedient.

Hut 8 has the interesting feature in the north-facing wall, described above and interpreted as a kiln flue by Paterson. This would work as a streamlined aperture, which would gather the north wind and vent the south (figure 5.18). With this aperture open to a north wind and with the southerly door/porch open, the through draft would create suitable winnowing conditions, similar to that at the Corrigal Farm grain barn. The porch, added to Hut 8 at a later date, indicates a later modification, perhaps because of the strong winds on Orkney. The built in wall, again added at a later date, might have been done to create a relatively draught free area in which to sit, making and mending tools. A building such as Hut 8 at Skara Brae where tools are made and repaired and where the grain may be stored, winnowed, malted and dried in a kiln should perhaps not be called a 'workshop' but rather a storage area and grain barn.

Figure 5.18
A good view of the kiln flue, on the seaward side of Hut 8.

Hut 7, Skara Brae

Standing on the natural sand, Hut 7 was apparently in use throughout the life of the settlement. The interior is well furnished. There is a stone 'dresser', three floor boxes of stone slabs sealed with clay, 'box beds' or 'pens' as Childe referred to them and a central hearth with a stone seat beside it. There is a stone-built plinth to the left of the door when entering that has an area of floor space enclosed by slabs in front of it. The long slab to the side of the hearth was not part

of the furnishings. It was a roof-supporting pillar that crushed "a very large pot with a decorated rim" that had been standing by the hearth when the roof fell in (Childe and Paterson 1929:259).

Artefacts found within the hut include small whalebone cups containing haematite pigment, stone mortars, various 'cooking pots' and several large Grooved Ware vessels. Pen D contained the skull of an ox (figure 5.19). The foundation burials of two old women were found underneath Pen Y and inscribed marks were noted on the front of the slab of this Pen (Childe and Paterson 1929:252-258). So far these inscribed marks are not understood nor have they been interpreted.

Hut 7 is the most low-lying of all the huts and Childe vividly described the lowest floor levels in this hut as being "... a slimy mass, having very much the consistency of blancmange. It consisted of saturated sand merging into the red clay of the floor, and containing, in suspension, broken bones, lost artefacts and all sorts of refuse. In this glutinous mass a multitude of large stones, mostly broken, were lying about in disorder, forming unstable and slippery islands on which one was glad to stand as refuges from the surrounding morass" (ibid: 250). Childe interpreted this as evidence that the inhabitants had lived in filthy and squalid conditions, but it was only this one hut that was described as "chaotic and disgusting" and "a morass of filth" (ibid: 259). Flooding could be the cause of such conditions, given the low level of the hut compared with the rest of the settlement. Another explanation might be that the hut was used over a long period of time for the consumption of ale.

A large highly decorated pot had stood by the central hearth. It was apparently smashed by the falling pillar when the roof collapsed. A fermenting wort needs to be kept warm. Had this pot been full of fermenting wort when it was smashed, which is likely given its position by the warm hearth, then the floor would have been flooded with sweet sticky wort or the fermented ale. If reeds had been on the floor during the life time of the hut, then the spilt ale and reeds mixture would decay as a disgusting mess, with lost items in suspension, exactly as described by Childe.

A finished and fermented ale would need to be transferred from one vessel to a clean one for storage. One way of doing this is to use gravity and, perhaps, reed tubes (figure 1.16). The pot would need to stand on the 'dresser' or perhaps on the stone plinth and the ale could then be syphoned into a suitable vessel standing on the ground. Reeds were available on Neolithic Orkney and would probably work as syphon tubes. Alternative methods could include holes in the pots or pouring the ale from one vessel into another, a tough, heavy and difficult task, or using smaller vessels to transfer the liquid. The transference of wort or ale from one vessel to another is a messy business. Accidents often occur and the sticky fermented ale is nearly always spilt on the floor. This might explain the disgusting state of the floor and the lack of a comparable mess on floors of other huts. Although all brewing equipment needs to be kept scrupulously clean, the consumption of the finished product can be messy.

Unique to this hut and situated to the left of the door is a large stone-built plinth, with an area before it enclosed by stone slabs. This would be suitable for the transfer of liquids

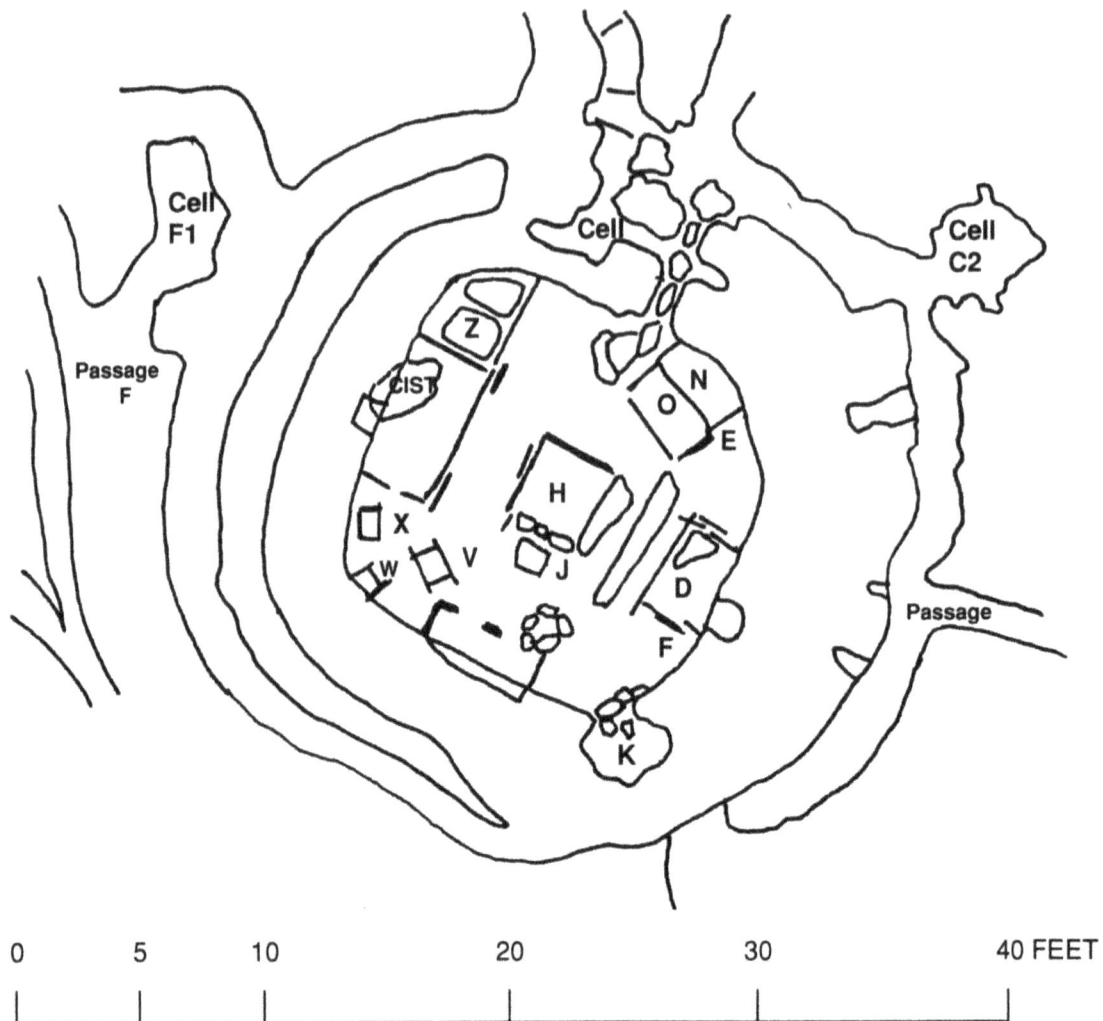

Childe's descriptions of contents of Hut 7

D – Pen containing an ox skull and green sludgy deposits. Tusk pendants, a bone scapula and pins were beside the pen.

E – Small whalebone cup made from the vertebra of a whale.

F – Little stone mortar with remains of a ceramic pot beside it.

G – Stone 'dresser.' The top shelf was empty; a Grooved Ware pot was on the middle shelf.

H – Central stone kerbed hearth.

J – A stone seat by the fire.

K – Cell in which a large Grooved Ware pot was found and a cache of bone beads and pendants.

N – Stone built plinth, with an area (**O**) in front enclosed by stone slabs, set on edge.

P – Circular pit surrounded by stones. Pit was 1ft 4in deep and filled with green sludge. A piece of Haematite was found at the bottom of the pit.

V, W, X – Three slate or stone lined pits, 1 ft 7in to 1ft 9in deep, clay sealed. Two had ox ribs in and the other was empty. A Grooved ware pot stood nearby.

Y – Pen which according to Childe was very filthy with a huge quartzite pebble and a large Grooved ware pot. A double female foundation burial was beneath the stone slab within the pen.

Z – Pen, slate floored, with two large cooking pots containing animal bones. Also a basin made of cetaceous bone, a stone mortar and a small cup made from a whale vertebra used to hold red pigment made from ground Haematite.

Figure 5.19
Ground plan of Hut 7, Skara Brae, showing the location of internal furnishings (Childe 1929:253-260). The long stone slab by the fire was probably a roof support which crushed a large, ornately decorated Grooved Ware bucket that had been standing by the fire and which may have contained fermenting wort or a fermented ale. The cist contained the bones of two elderly women, perhaps a foundation burial. This image is traced from figure 5.15 (Childe 1931).

Figure 5.20
The plinth beside the door in Hut 7 with stone slabs in front.

Figure 5.21
Inside Hut 7, looking through the doorway towards Pit P.

by syphon, the spillage being contained within the area built before it (figure 5.20). A small 'ambry' or keeping place, about the size of a single stone, is built into its front face, the purpose of which is not clear. Without more evidence it is difficult to clearly attribute either a mundane or a special ritual activity to this plinth.

At various times in his description of the excavations, Childe refers to a "green slime" or "green sewage clay" that he noted to be in most of the drains. At the base of Wall Q this green slime reached a depth of up to about 20 inches (Childe 1930:34,41). He interpreted it as being the probable remains of human sewage, the main function in his opinion for the drainage system. Unfortunately there has been no scientific analysis of this substance, nor, apparently, have recent excavations come across it in any of the drainage systems of the more recently discovered Neolithic sites and domestic structures on Orkney (Richards pers comm).

There are no drains attached to Hut 7 and yet this 'curious green substance' was found within the Hut in Pit P, a circular stone-lined sump at the foot of the 'dresser'. It was also found in the base of Pen D (figure 5.21). Both of these green deposits went through the clay floor and into the underlying sand (Childe 1929:259). It is possible that Childe's interpretation of the green material as decayed sewage in the drains is correct. If so, then what was happening at the 'dresser' in Hut 7, to allow a thick deposit of this substance to leach through the clay floor into the sand? A more plausible explanation is that the green slime represents decayed sugar residues from spilt wort or ale within Hut 7.

Hut 4 and Hut 5, Skara Brae

Thick deposits of the "green slime" or "green sewage clay", as described by Childe, were found by wall 'Q' close to Huts 4, 5, 9 and 10 (figure 5.15) apparently representing some kind of discharge from those cells (Childe & Paterson 1929:253,259). Drains run under both these huts, apparently connecting cell 2 of Hut 5 and cell 1 of Hut 4 and continuing towards the cell of an earlier Hut 4. There is a huge 'sump' beneath Hut 5, the precise function of which is not at all clear. Another drain (Drain E) runs from Hut 5, under the main passage and out to sea (figure 5.15).

Any grain processing that took place within these huts would produce the washings of used pottery. Brewing or grain processing residues, waste liquid and sugars would wash into the drains and decompose, perhaps creating the green slime found in so many places at both Skara Brae and Rinyo by Gordon Childe. Future excavations at Neolithic Orcadian sites should look out for any of this green substance so that it can be properly analysed. The drains at Skara Brae present a confusing picture but one that merits much further investigation than this brief analysis.

Hut 5 may possibly have been used for the wet processing of barley mash, namely, obtaining a sweet wort or malt liquid by sparging the barley mash with hot water. The hut is furnished with floor boxes, box 'beds' and a central hearth. The floor boxes would function well as kettles or water heaters. Any water stored inside them could be easily heated up with hot stones and subsequently used for mashing or other purposes. Many huts at Skara Brae have these floor boxes, usually referred to as 'limpet boxes', that is, for the live storage of limpets to use as food or fishing bait. This does not seem a likely function. Limpets would soon become very smelly. They require sea water and regular daily tidal changes in order to survive. A more suitable function for the boxes would be as 'kettles' for a supply of hot or warm water. In Hut 7, ox ribs were found in two of the floor boxes (Childe 1929:253), so they could also be used for cooking meat.

Hut 4 has internal structural similarities to an animal barn, with vertical slabs set into the ground to delineate spaces. These are about the right size for animal stalls. The internal furnishings of Hut 4 are unlike the other huts and merit much further investigation. How big were the cattle at Skara Brae and would they have fitted through the entrance passage into Hut 4? These issues are not directly related to the manufacture of malt and ale and will be the subject of future research, together with a detailed analysis of the probable functions of Huts 1 and 2, which are more domestic in character than any of the huts discussed so far.

In summary, the architects, builders and inhabitants of Skara Brae had a clear design in mind when they arrived in their boats at the Orkney Islands and began building the huts for their community. Skara Brae has several levels of construction and development within it. In many respects it could be described as a settlement tell, albeit on a small scale, that is similar to those in the Near East. Skara Brae has complex levels and there are confusing inter-relationships between the buildings. The inhabitants here created and continually modified a practical and functional living space.

They needed somewhere safe and suitable to raise their children, to store, to cook and to prepare their food, to process their grain into malt and ale, to stall their animals and to provide shelter and warmth for all in both summer and winter. The Neolithic settlement at Skara Brae could be understood as being very like a traditional crofting community, with different activities being pursued depending on the time of year, the prevailing weather conditions and many other practical considerations. Making sweet malts and ale was only a part of a complex lifestyle.

Rinyo, Bigland Round, Rousay

The settlement at Rinyo, on Rousay was excavated by Gordon Childe in the late 1930s and the early 1940s (Childe 1938, 1946-8). The walls of the buildings did not survive as well as at Skara Brae and Knap of Howar and the village is now overgrown. Nothing remains visible above ground today. Childe took many photographs and accurate plans and drawings were also made. These are now the main means of investigation and interpretation of this site, unless it is re-excavated in the future. According to Childe, (1938) there were at least 7 houses and 2 major phases of rebuilding. Childe excavated 4 of the 'chambers' and thought that the village may have been significantly larger than Skara Brae (figure 5.19).

Grooved Ware pottery was used at this settlement, with large flat based vessels that were comparable in size to those at Skara Brae and similar to those at Knap of Howar. Stone pot lids measuring up to 22 inches in diameter were also found. Childe noted that hunting equipment and fishing tackle were "notably absent" from the site, although it is possible that the acid soil may have destroyed such bone tools and implements. Stone drains, some lined with hazel bark, were found at Rinyo. Also of interest to grain processing activities such as malting and mashing are the many ovens and hearths that Childe and his team uncovered during excavations. A particularly large oven and hearth was found in Chamber C and a clay oven base was found under the floor of Chamber A (figures 5.25, 5.26).

All the necessities for making ale from the grain are evident at Rinyo. There are stone 'dressers', hearths, suitable pottery vessels and a drainage system with evidence of the green sludgy deposits (Childe 1946-8:24). Many stone boxes are set into the ground that, as discussed earlier, could have provided hot water for a multitude of purposes. The immersion of fire-heated stones would suffice for this. It would seem that life at Rinyo was very like that at Skara Brae. It is worth noting that Rousay, of all the Orkney Islands, has the largest number of stalled and chambered tombs, in total, 15. Was the settlement at Rinyo the only one on the island, or are there more yet to be discovered? Did the settlement have a special significance, perhaps to do with burial and mortuary practices? The site merits further investigation.

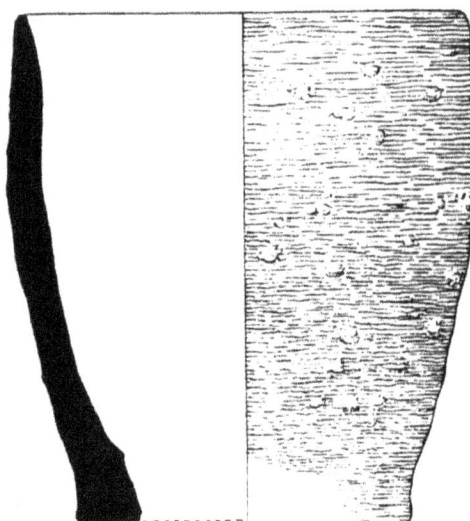

Figure 5.22
A 'large' pottery bucket from Rinyo (after Childe 1938:24).

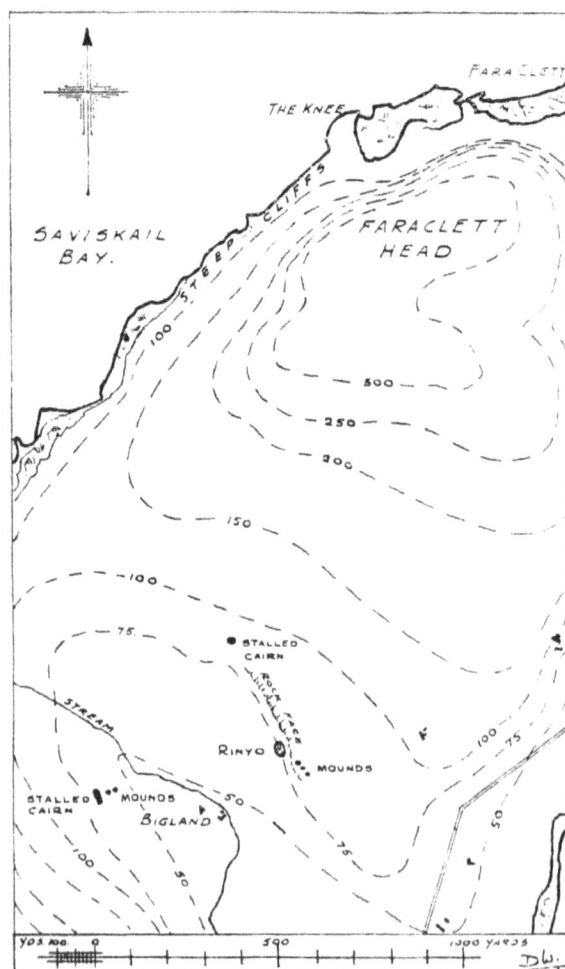

Figure 5.23
Map of Rinyo on Rousay (after Childe 1938:8).

Proc. Soc. Ant. Scot.]

RINYO, ROUSAY, ORKNEY.

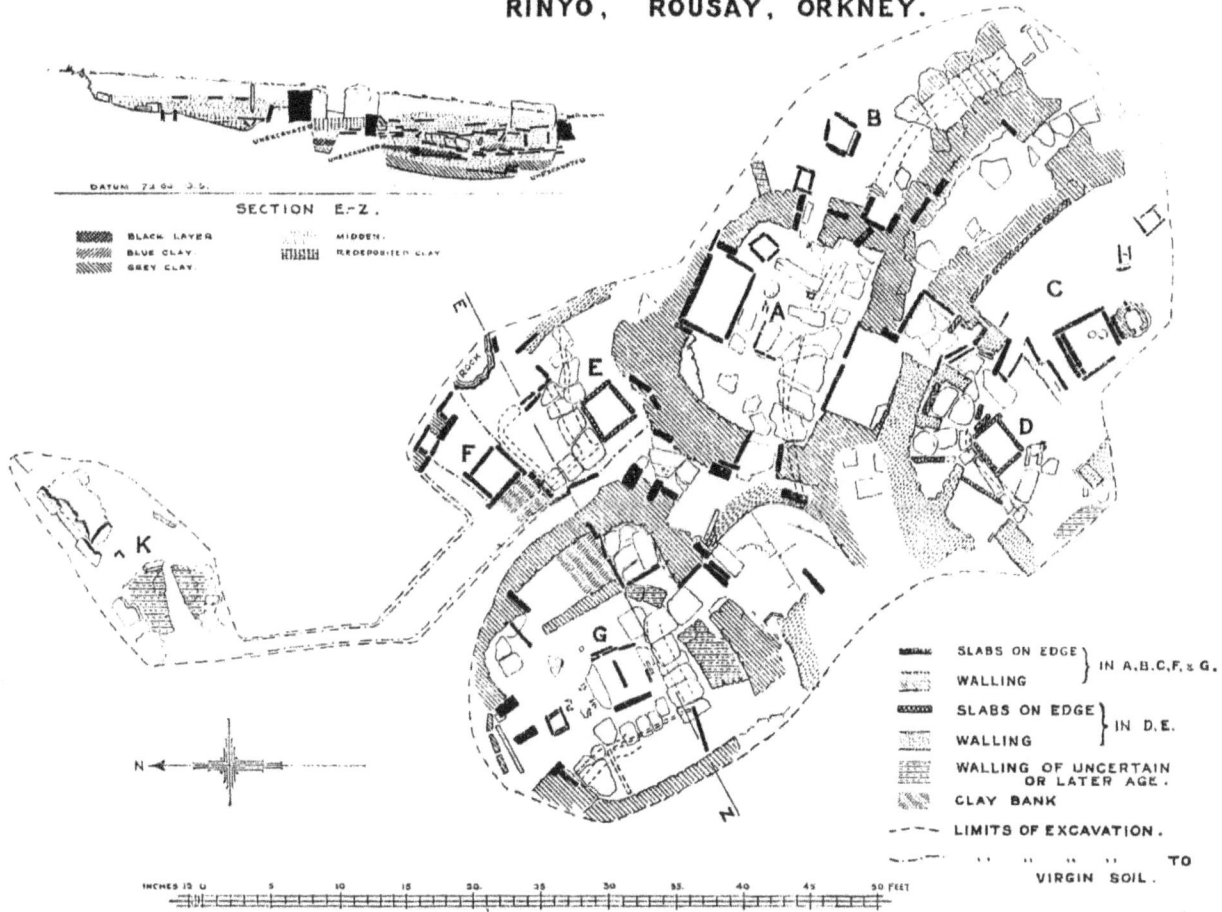

[VOL. LXXXI.

SECTION E-Z.

BLACK LAYER
BLUE CLAY
GREY CLAY

MIDDEN.
REDEPOSITED CLAY

A   B   C   D   E   F   G   H   K

SLABS ON EDGE } IN A,B,C,F,& G.
WALLING
SLABS ON EDGE } IN D,E.
WALLING
WALLING OF UNCERTAIN
      OR LATER AGE.
CLAY BANK
- - - LIMITS OF EXCAVATION.
—— "   "   "   TO
            VIRGIN SOIL.

N

INCHES 12 0   5   10   15   20   25   30   35   40   45   50 FEET

Figure 5.24
Ground plan of Rinyo, showing huts excavated and the drainage system, as dotted lines beneath huts B, A, G and E. The drains are stone built with a hazel bark lining (Childe 1946:17). Childe reported finding green slime as at Skara Brae.

Figure 5.25 (left)
The oven and hearth in Chamber C, Rinyo, Rousay. Scale is indicated by the circular tape measure, placed centrally and about 6 inches in diameter. The hearth is large, about 30 inches wide. There appears to be an oven-like structure beside it.

Figure 5.26 (right)
The oven base found under the floor of Chamber A, indicating the existence of earlier buildings at Rinyo. Scale is indicated by the tape measure.
(Childe 1938, Plates VI and IX).

Other Orcadian settlements

A number of other settlements are known on the Orkney Islands but, unfortunately, for most of these sites the excavation reports are yet to be published in full. Neolithic settlements have been excavated on the island of Sanday, specifically at Pool, Tofts Ness and Stove Bay. All had both Grooved Ware and Unstan Ware pottery and buildings that appear very similar to those at Skara Brae, Rinyo and Barnhouse. The site at Stove Bay is being rapidly destroyed by the action of the sea.

The settlement at Pool had at least 14 houses and a complex series of midden deposits (Hunter & McSween 1991). The pottery sequence discovered at Pool is particularly interesting, since it appears to provide a "chronological and material sequence, which transcends the Unstan-Grooved Ware divide" and a very large number of vessels, almost 2000, appear to be represented by the pottery sherds found there (Bond 1995:126/7).

Tofts Ness seems to have been a small settlement containing one or two houses and an assemblage of Grooved Ware. Stove Bay is rapidly eroding out of the cliffs and is being slowly destroyed by the sea. Walls, house floors and stone kerbed hearths are visible in the cliff face (Bond 1995).

A much larger Neolithic settlement on the north coast of Westray known as Links of Noltland is also being eroded and destroyed by the Atlantic Ocean. Although discovered by Petrie in 1866, it was not excavated until the 1970s (Clarke 1978). It appears to have been about four times larger than Skara Brae (Fraser 1983:146). Full analysis of these sites must await publication of excavation reports.

2. Women in Neolithic Orkney

In Iron Age, Viking and Medieval times women have been generally and frequently responsible for the growing and tending of crops and for most domestic food processing activities, which would include the making of malt, malt products and ale (Bennett 1996, Davidson 1998). It is a strong possibility that women in Neolithic communities, such as those on Orkney, were the principal grain cultivators, processors, maltsters and brewers. They would have possessed the necessary skills, the knowledge and the special secrets and techniques of transforming the harvested grain into sweet malts and ale.

This would have made the role of women in Neolithic communities an important and varied one, involving activities like childcare, crop cultivation, food gathering and processing. Their knowledge of the use and properties of plants and herbs must have been extensive. To give an example, pregnancy and childbirth are areas of experience that are exclusive to women and that frequently require special skill, knowledge and experience in the use of particular herbs to aid and assist at these times. Hilda Ellis Davidson (1998:8) has pointed out the diverse roles of women in prehistoric and early historic times, with their use of herbs in healing rituals being an important aspect of their specialised knowledge.

Two arthritic female skeletons, perhaps ritual foundation burials, were discovered under Hut 7 during Childe's excavations at Skara Brae. This might be an indication of the importance of these particular women within the community. Perhaps they were the healers, wise women,

maltsters, brewers or spiritual leaders. Given Childe's description of the messy state of the floor of Hut 7, as discussed earlier, it is more likely that activities within Hut 7 involved consumption rituals rather than rituals involving the processing of grain.

A strong indicator of special ritual activity at Skara Brae is the presence of burnished or polished Haematite. This is found in numerous places including Pit P in Hut 7, Huts 1 and 3 (Petrie 1867:210), in other Huts and in the middens. As demonstrated by the Orcadian artist Arlene Isbister at the Neolithic fair that was held on Orkney in conjunction with the academic conference in September 1998, Haematite is a source of red/brown or orange/yellow pigment. It can be used to paint and decorate pottery, stone, leather and also the human body. It has many qualities that make it a potentially magical substance (Isbister 2000). It is a black stone that produces red, the colour of blood. Interestingly, it even has a blood-like taste because of the high iron content of the ore. Haematite is located in only one or two places on Hoy, where it is found in "thin veins and fractures...with black compacted nodular lumps and fist-sized, kidney shaped masses" (ibid). Arlene Isbister's experiments show that it is possible to get colour from the ore very easily - simply rubbing nodules with water on a hard smooth stone surface produces pigment in shades varying from blood red to red-brown, orange and yellow. Working on a hot stone slab or on a cold stone slab produces variations of colour. Different faces of the same nodule of Haematite can produce different shades of red, yellow, orange or brown.

Haematite has been found during excavations at most of the other Neolithic Orcadian sites and also at Çatal Hüyük, Turkey. Mellaart (1967:149-50) argued that it symbolised blood and life, having the power to protect against evil forces. Ritchie (1995:34) has argued that its discovery in Hut 7 might point to confinement-related interpretations or to other rituals and rites of passage, such as initiation rites, childbirth and menstruation. Barbara Walker (1988, referred to in Isbister 2000) believes that Haematite represented blood, both medically and mystically. It has chemical properties that arrest bleeding and so it might have been used both practically and ritually (Isbister 2000). The Orcadian Neolithic culture appears to have had a powerful female element that is neither easy to define nor to compartmentalise. Rituals undertaken in Hut 7 were probably complex events, involving a number of different activities, including childbirth, the use of Haematite, the consumption of special brews or potions and much more besides.

There is a burial cist that contained a female skeleton situated beneath Building 2 at Barnhouse (Richards 1992). This building was apparently related to aspects of grain processing activity. Perhaps a similar practice of honouring an important wise woman of the community is indicated here, as a foundation burial in a building within a settlement that is situated only half a mile from the Stones of Stenness. Both Hut 7 and Building 2 were in use throughout the lifetime of the settlements, perhaps an indication of the special and unchanging use of the buildings and the importance of those women buried beneath them.

The Neolithic communities at Skara Brae, Barnhouse and other Orcadian sites had a material culture suitable for making sweet malts and ale from the barley and wheat that they grew. They had pottery vessels, specially built drains and suitably designed buildings. The transformation of grain into ale involves a specific set of domestically-based ritual activities and processes. The barley or wheat grain must be planted, tended and cultivated before it can be harvested, threshed, winnowed, malted, mashed, sparged and finally fermented into an intoxicating drink, perhaps with special or medicinal herbal additives being added during the boil. The techniques for each stage are quite specific. The order in which the tasks are done cannot be altered and there is much that can go wrong at different stages of the whole process. Fermentation was believed to be a magical event until recent historical times when, as discussed in Chapter One, Louis Pasteur and John Tyndall discovered the scientific and biochemical explanations.

The magical properties of the barley grain may have made it a special or even a sacred crop during the Neolithic. Women were very likely to have been responsible for the sowing, cultivation, harvesting and complex processing of the grain. Euan MacKie has proposed that Skara Brae was "a settlement of an elite of wise men or professional priest astronomers, comparable to those ... at the great henge sites of the south, like Durrington Walls" (MacKie 1997:339). His interpretation of only male 'priest astronomers' at Skara Brae is supported neither by the archaeological evidence of the female foundation burials that are discussed above nor by the analysis of the function of some of the Huts of the village. From this evidence it might be argued that women were the spiritual leaders during the Orcadian Neolithic. However, a more realistic and valid explanation would include both males and females as spiritual leaders of the community, each with specific areas of influence.

MacKie notes the drains that run under the village of Skara Brae and also the separate building, Hut 8, which he interprets as a combined workshop and cookhouse. He concludes that these factors made Skara Brae much more than a "simple peasant settlement" (ibid 339). That Skara Brae was a special village is an interpretation open to debate. Some potential reasons for the drains and the probable functions of some of the buildings have been discussed earlier in this chapter. Drains are a necessary and specifically designed feature. They were deliberately constructed to facilitate the processing of grain into malts and ale, an activity that may have been the responsibility of the women of the community rather than the men.

MacKie points out that "any picture of the Grooved Ware period in Orkney must notice the evidence for unusually elaborate ceremonial activity" (MacKie 1997:339). He notes that Renfrew believes Orkney to have been a very significant place in Neolithic Britain. It was a place with "a remarkably powerful body of religious beliefs, with accompanying ritual observances" (Renfrew 1990:256). Barnhouse, situated by the Stones of Stenness and within a mile of the Ring of Brogar, was also a ceremonial and ritually significant place, housing a group of 'a religious elite' that are believed by MacKie to have been male (MacKie 1997:339). This preliminary research into malting, mashing and brewing technologies indicates that some members of this 'religious elite' were female.

It appears that the activities involved in the manufacture of ale may have had a powerful ritual significance to Neolithic people. The women who may have made the ale would have held positions of power, influence and status. They had more responsibilities than simply malting and brewing, being the healers and leaders within the community and having an influential social role as well as a ritual one.

The Grooved Ware Culture was not confined to Neolithic Orkney. Grooved Ware pottery has been found, apparently smashed and deliberately deposited in pits, for example, at the ceremonial site of Balfarg/Balbirnie in the Tay valley on the east coast of Scotland. Organic residues on this pottery included burned cereal mash as well as pollen from herbs, including Henbane and deadly Nightshade (Barclay et al 1993). Cereal based residues have also been identified on pottery sherds from Machrie Moor Stone Circle, Arran, and on Neolithic sherds found at Kinloch Bay, Rhum. By the early 3rd millennium BC, huge quantities of Grooved Ware pottery buckets and vessels were being used at Durrington Walls, an enormous henge with timber circles and rectangular buildings that has only been partially excavated and that is located nearby Stonehenge and very close to the river. These sites are discussed in the next chapter, which investigates the Grooved Ware culture and the evidence for the transformation of grain into sweet malts and ale in Neolithic mainland Britain.

CHAPTER SIX
The Grooved Ware Culture in Neolithic Britain

1. The transition to agriculture

Elements of both the European and the Orcadian Neolithic cultures are evident in the archaeological record of mainland Britain. Monuments were constructed in the form of stone and timber circles, tombs, long mounds, causewayed enclosures and passage graves. Ceramics technology was introduced to Britain at the same time as the practice of grain cultivation and processing. Domestic animals were also introduced in the 4th millennium BC (Ashmore 1996). Ritual activity and feasting appear to have been very important aspects of early Neolithic life and there are numerous ritual sites and ceremonial centres. Examples of stone circles include Stonehenge and Avebury in the south of mainland Britain, Callanais in the Outer Hebrides, Machrie Moor on the Isle of Arran and Arbor Low and Stanton Moor in Derbyshire to name but a few (figure 6.1). There are literally hundreds more henges and extant stone circles as well as the evidence for many timber circles throughout the British Isles (Gibson 1998). These were important places that were constructed by the local Neolithic groups and communities, places of ritual activity where people may have met regularly for feasting, for fairs, perhaps, and for many other activities and reasons.

To date, longhouses of the Linearbandkeramik style have not yet been found in Britain although timber buildings of a variety of sizes and shapes are evident in the British Neolithic. At Balbridie, Kincardine, Scotland, (Fairweather & Ralston 1993) excavations revealed the postholes of a huge timber hall that, according to radiocarbon dates, was constructed during the 4th millennium BC and destroyed by fire. Large amounts of carbonised grain were found there, suggesting its use as the community storage or processing centre. Wheat has been found in far greater amounts than barley. 'Spent grain', the term used for what is left after the malt sugars have been sparged or washed out of the barley mash, may have been fed to the animals. Unsparged barley mash is far too sweet to be eaten by domesticated animals. However, once the sugars have been washed out (or sparged) it makes excellent cattle feed and it is still used today for this purpose. This may explain the far fewer finds of barley grain on the site compared with wheat.

Although pedestalled bowls of the style of the middle Neolithic TRB cultures have not been found in British passage graves or at ritual sites, Grooved Ware bucket-shaped pottery and Unstan Ware, which is wide, shallow bowls, are frequently found on Orkney and throughout mainland Britain at both ritual and settlement sites dated to the 4th and 3rd millennia BC. Both styles are very similar to the Drouwen TRB West pottery (figure 4.12)

Hunting, gathering and fishing groups inhabited the Western Isles, the Hebrides and mainland Britain during the 7th, 6th and 5th millennia BC (Wickham-Jones 1990). Mesolithic sites are fewer in number than Neolithic sites but this should not be understood to reflect population numbers but rather the great difficulty in discovering and identifying Mesolithic habitation sites. The acceptance and the adoption

of grain cultivation and processing at the beginning of the 4th millennium BC was swifter in Britain than had occurred on the Northern coasts of Europe. In both areas, fishing, hunting and gathering remained important subsistence strategies alongside the new food technologies of grain cultivation and processing (Armit & Finlayson 1992). There are indications of the continued use of shell middens from the Mesolithic right through to the late Bronze Age (Renfrew J. 1985:10). So why, then, did Mesolithic cultures of the British Isles wish to begin to sow, cultivate, harvest and process grain? The answer lies in some of the potential products of grain.

As has been discussed in earlier chapters, the pottery styles of the TRB, Linearbandkeramik and Orcadian Neolithic indicate that grain processing included the manufacture of a liquid product such as ale. This was probably consumed at private feasts, public festivals, celebrations and work-party feasts (Hayden 1996). Similar grain processing activity appears to have been practised by Neolithic cultures on mainland Britain and there is some convincing archaeological evidence. They had suitable pottery vessels for mashing in the form of wide, shallow pottery bowls. For fermentation, there were large, deep flat-based pottery buckets. Suitable buildings existed, constructed of stone on the Orkney Islands and of timber on the mainland and elsewhere, for the storage and dry processing of grain into malt, the wet extraction of malt sugars and the fermentation of those malt liquids into ale.

This chapter assesses some of the archaeological evidence for malting, mashing and fermentation in mainland Britain by the Grooved Ware culture in the early Neolithic. The evidence of suitable pottery vessels, such as ceramic bowls and Grooved Ware vessels will be discussed. Organic residue evidence on pottery sherds will be evaluated and, the evidence of timber buildings for the storage and processing of grain will be examined.

2. Suitable pottery vessels for ale

It is essential to look at the size, shape and potential function of pottery as well as at its classification into a particular decorative style. Ceramic bowls, frequently associated with timber structures, have been identified as having "a particular social importance in the earliest Neolithic of the British Isles" (Kinnes 1985:20). Early Neolithic bowls were round based and were probably used for cooking food. They are, of course, quite suitable for mashing the malt in the warm ashes of a fire, as illustrated in Chapter One. In the earliest Neolithic of Britain the manufacture of ceramics was a brand new technology, introduced to the island in association with completely new food resources, namely cereal grains and cattle. The grain had to be prepared in a number of special new ways and it seems to be assumed in most of the current archaeological literature that the potential products of grain were only flour, bread, porridge or gruel.

Grain was malted, and mashed to produce a sweet liquid that was fermented into an alcoholic drink. These new food preparation techniques required special knowledge. Prior to the discovery of making malt from barley, the only sources of sugars available for fermentation were honey or fruits (Crane 1983). It is likely that women possessed the skill of cultivating the new food resource and that they passed on their special knowledge and the secrets of the new techniques.

Orkney Islands

Calanais

Raigmore

Balbridie

Balfarg / Balbirnie

Kinloch Bay, Rhum

Machrie Moor, Arran

Ballynagilly

Ronaldsway,
Isle of Man

Ballyglass

Knowth

Lismore Fields
Arbor Low

Llandegai

Lough Gur

Avebury

Stonehenge Durrington Walls

0   200km

Figure 6.1
Map of the British Isles, Ireland, the Western Isles and Orkney, showing the approximate location of sites discussed in the text.

Grooved Ware pottery was first described as being 'bucket-shaped' by Stuart Piggott (1931:73). He originally named this pottery style 'Rinyo-Clacton Ware' because of its presence throughout the whole of the British Isles during the early Neolithic (Piggott 1936). The Grooved Ware pottery tradition apparently originated on the Orkney Islands in the mid 4th millennium BC and the use of it spread within several generations to eastern and southern Britain. By c2800 BC it was being used at and around Stonehenge. A very large amount of Grooved Ware sherds have been found at Durrington Walls (Castledon 1990:85, Souden 1997:57). Grooved Ware has strong connections with ritual activity and it is frequently found smashed and deliberately buried at ceremonial and ritual sites. For example, some was found in the central hearth of the Stones of Stenness, Orkney (Ritchie 1976). As noted in earlier chapters, Grooved Ware is an ideal size and shape both for drinking from and for the storage and fermentation of several gallons of ale.

Grooved Ware sherds representing approximately 35 vessels were found at the site of a "ritual rather than domestic" rectangular timber building at Stoneyfield, Raigmore, Inverness, Scotland. The vessels varied in size from "a small cup of 140 mm (P6) to very large vessels up to 460 mm in diameter (P23). Wall thickness was between 8 and 15mm" (Simpson 1996:66). This is an assemblage for a liquid product, with the largest vessels suitable for fermentation or storage purposes; the smaller ones would function as drinking vessels. A large number of the Grooved Ware sherds had been deposited in pits. One interpretation of Raigmore, on the basis of the evidence and research presented so far, is as a grain barn or a central processing centre for grain.

3. Organic residues

There have been a number of discoveries and analyses of organic residues on pottery sherds found in Neolithic contexts over the last ten years that support the concept of grain processing for malt and ale. The analysis of residues on pottery provides evidence for the probable use of the vessel. Residues have been found on Grooved Ware pottery sherds from the 4th millennium BC ceremonial site at Balfarg, Fife, Scotland (Barclay et al 1993) and also from pits by the stone circles at Machrie Moor on the Isle of Arran (Haggerty 1991). Analyses of these residues show a consistent association between barley processing activity and flat-based pottery and large Grooved Ware vessels. The findings of barley residues in association with Grooved Ware pottery from Barnhouse, Orkney, have already been discussed in Chapter Five (Jones 1999). The residue analyses from finds at Balfarg and Machrie Moor are discussed here.

Balfarg/Balbirnie, Scotland

The fertile valley of the river Tay has been identified as one of the few patches of cultivable land in Strathearn. It appears to have been cleared of trees, if not by the builders of the henge then by earlier communities. Pollen analysis of a radiocarbon dated core taken from North Mains, Strathallan, dates the first grain cultivation in this area to c3600 BC (Hulme and Sherrif 1985), some of the earliest in Scotland and England. Recent excavations carried out between 1970 and 1985 by Graham Ritchie, Roger Mercer, Gordon Barclay and others revealed "the remains of a great prehistoric religious centre, in use from before 4000 BC until after 2000 BC" (Barclay 1996:4). This is the ceremonial site at Balfarg/Balbirnie situated within the bounds of the Rivers Leven and Eden in the Tay Valley, Scotland. Here, there is evidence of rectangular timber buildings or enclosures that are interpreted as excarnation centres, in use during the Neolithic. A stone circle, henge, ring cairns, cremations and Bronze Age cist burials have also been found in the area.

The pottery assemblage included carinated bowls that were interpreted as having been deliberately broken and then deliberately deposited in pits. A single barley grain was noted within the fabric of one of these bowls (P23) (Barclay et al 1993:71). Christopher Tilley has recently suggested that the inclusion of grains into the fabric of early Neolithic pottery vessels may have been a deliberate and a symbolic act rather than just an accidental occurrence, thus creating "a direct symbolic link.... between pottery, cooking, grain, fertility and ancestral powers" (Tilley 1996:189). His view supports the idea presented earlier in this thesis that the barley grain itself had important ritual and symbolic value to Neolithic people.

At Balfarg, large Grooved Ware vessels had been smashed and the sherds deposited in pits, close by one of the two rectangular timber structures (Barclay et al 1993:183). These timber structures were interpreted as being mortuary enclosures not roofed buildings, hence the context of the pottery sherds is ritual rather than domestic and more to do with consumption rituals rather than the ritual of the manufacturing process. Burned organic material was noticed on sherds from two of the largest vessels in this assemblage, pots numbered P63 and P64 (figure 6.2). This organic material was analysed by Brian Moffatt (in Barclay et al 1993:108*ff*). His results have proved to be controversial. The descriptions of the residues provide some interesting evidence for ritual activities in the Neolithic and for the ritual consumption of ale in a funerary context.

Three categories of burned organic material were defined: amorphous and burned material, amorphous granular and burned and burned cereal mash. Both barley and oats were identified within the last category, having been "thoroughly ground down" thus making precise taxonomic identification difficult. These macro plant remains were described as a "cereal based preparation" that was "coarse and crude" (Moffatt in Barclay et al 1993:108,109). This is interpreted as having been "coarse porridge" but, with reference to Chapter One of this thesis, it is much more likely that these are the residues from brewing, being the sediment that always settles out to the bottom of vessels used either to store or to ferment ale (figure 1.12). There are striking similarities between the descriptions of the residues found at Balfarg and the appearance of barley residues that are obtained when making an ale or beer. Brian Moffatt notes the "incomplete process of homogenisation" and also that the "pollen and seed fragments were fairly well intermixed" (ibid: 109). Such descriptions are entirely consistent with the residues obtained in the mashing and fermenting experiments that are described in Chapter One of this thesis.

Pollen grains and macro plant material as well as minute amounts of beeswax were identified on 15 of the 31 samples from the Balfarg Grooved Ware sherds (ibid 109), the

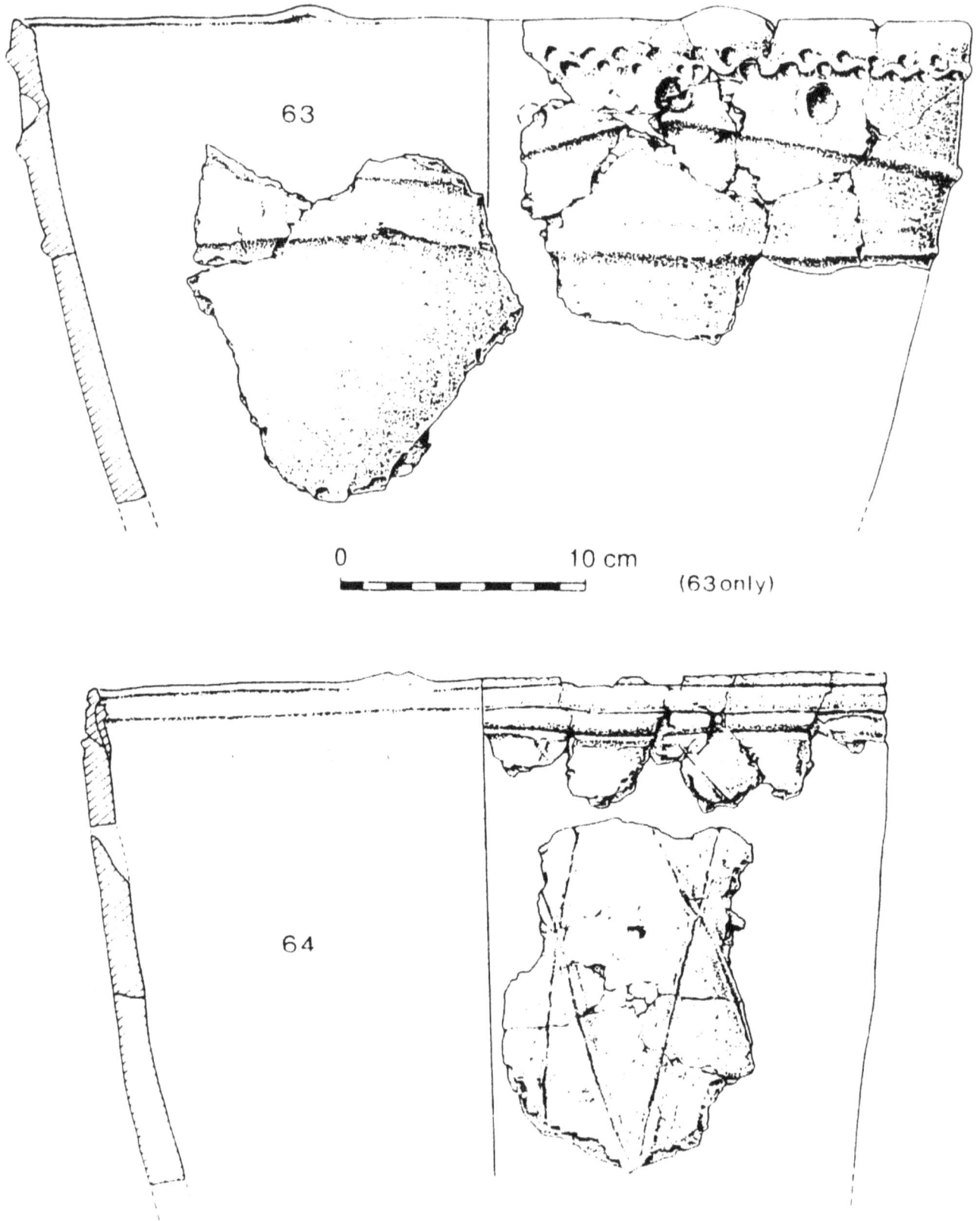

63

0   10 cm

(63 only)

64

Figure 6.2
Reconstruction drawings of the two largest pottery vessels from Balfarg ceremonial site. Organic residues consisting of burned cereal mash, pollen and henbane seeds were identified on sherds from these pots, a good indication that the pot had contained a fermenting wort with possible hallucinogenic additives. These Grooved ware sherds were found close by a timber mortuary structure and apparently, they had been deliberately deposited in pits (after Barclay et al 1993:102,103).

beeswax probably being used to seal the vessel to prevent the leaching out of contents through the otherwise porous fabric of the pot. The pollen and macro plant remains from Meadowsweet *(Filipendula ulmaria)* indicates the addition of whole flowerheads to the brew, probably during the boil to act as flavouring and as a preservative (see Chapter One). There were also pollen grains from fat hen, cabbage and mustards. A single pollen grain of Deadly Nightshade *(Atropa belladonna)* was found as well as a number of seeds of Henbane *(Hyoscyamus niger)* and Hemlock *(Conium Maculatum)*.

Henbane, Hemlock and Deadly Nightshade are highly dangerous plants. Hemlock is poisonous. It acts as both a sedative and as an anesthetic. A decoction of the herb was routinely used by the Ancient Greeks as a means of execution. This was the herb consumed by Socrates for his suicide. Henbane and Deadly Nightshade contain phytochemical alkaloids which, in small quantities, depress motor function, and act as painkillers, inducing sleep and hallucinations (see Chapter Two). Large amounts of henbane are poisonous inducing dementia, paralysis and ultimately death. There is no known antidote. The deposit encrusted on the outer surface of sherd P63 was examined in more detail and it was noted that some of the seeds of black Henbane, which are described as "extremely robust and resilient" appeared to have been "rendered mechanically" (ibid:109). This perhaps implies their deliberate preparation for a medicinal or, more likely, a special ritual purpose.

Brian Moffatt lists the symptoms of henbane poisoning as including "blurred vision, dilated pupils, rapid heartbeat, dizziness, nausea, headache, euphoria and hallucinations" (ibid:110). There is a strong possibility that the plant was used deliberately to induce intoxication, hallucinations or perhaps even death. The excavators refer to the deposited Grooved Ware sherds as having contained "ritually charged material" (ibid:185). The ritual consumption of Henbane Ale in the context of a Viking chieftain's funeral, as described by Ibn Fadlan in the 1st century AD, was noted earlier (Chapter Two) and it is possible that similar ritual practices may have taken place at the early Neolithic ceremonial site at Balfarg.

Machrie Moor, Isle of Arran

At the stone circles at Machrie Moor on the Isle of Arran off the West Coast of Scotland, organic residues were found on both internal and external surfaces of pottery sherds. Unfortunately, it was not possible to assign the residues to specific pottery types (Moffatt in Haggerty 1991:91). Both Grimston/Lyles Hill bowls and Grooved Ware buckets were represented in the pottery assemblage. The bowls were found as sherds deposited in pits that were associated with the timber circles that pre-date the existing stone circles. Radiocarbon dates of c2870 were obtained from charcoal samples taken from the pits in which the bowl sherds were found (Haggerty 1991:57). The Grooved Ware sherds were found in postholes and charcoal gave radiocarbon dates of between c2520 and c2030 BC (ibid: 58). The pottery was associated with the timber circles, being found in postholes of the main circle and central setting and also in soil disturbed by the construction of the later stone circle.

Analysis of the organic residues left on these pot

sherds revealed clumped, immature flower pollen, probably from broken up flowerheads that had been picked specifically for the purpose, as well as birch/pine sap or resin, hazel (in the form of nuts and buds) and cereal pollen. The interpretation was that the organic residues represented inclusions in a mead-type drink, but it was not possible to elucidate a recipe from the organic residues analysed nor to re-create the concoction (ibid: 91).

Activity at the site included the cultivation of barley, evidenced by small quantities of barley pollen found in some of the extensive ardmarks in and around the ritual site. Contemporary with the ardmarks were postholes that indicated fences or stake lines. These were radiocarbon dated to c1890 BC. Algal spoors of seaweed indicate the manuring of the cultivated land. Impressed Ware and Beaker sherds were also found (ibid:86). Apparently, barley was grown within the bounds of the sacred precinct. Given the location of the ard marks over the whole of the ceremonial site, even within the timber and stone circles themselves, it seems that the barley may have been a sacred or a ritual crop.

Similar ritual cultivation of barley appears to have taken place close by the henge and stone circle at Moncrieffe Hill, Strathearn, Scotland (Stewart et al 1985). The excavator comments that "the builders of the henge were prepared to respect certain ritual practices: they sited the monument near water, they gave it the required NE orientation – yet they were prepared to cultivate grain within a stone's throw of a sacred precinct" (ibid: 141). Grain cultivation and processing may have been a special activity in the early Neolithic.

Recent research shows that grain was not used as a major food resource in the British Neolithic. Mike Richards (1996) analysed samples of human bone using stable isotope analysis as a method. His results indicate that meat was the main protein resource of the Neolithic. He concludes that grain was a crop grown for ritual purposes. The findings at Machrie Moor and Moncrieffe Hill support this view.

The Neolithic ceremonial sites at Balfarg/Balbirnie and Machrie Moor provide convincing evidence for the ritual manufacture and consumption of fermented alcoholic drinks in the 4th and 3rd millennia BC. At Balfarg, the organic residues found on the Grooved Ware are consistent with the remains of ale. Powerful psychoactive drugs, Henbane, Hemlock and Deadly Nightshade were possibly added to the brew.

Evidence presented in earlier chapters points to the female element in the cultivation and processing of grain and in the manufacture of malt and ale. Women would have been responsible, in mainland Neolithic Britain as in the Ancient Near East and on Orkney, for such tasks and therefore women probably had a leading role in some of the ceremonies that were associated with ritual monuments.

The evidence for Neolithic settlements, apart from the stone buildings of Orkney, already discussed in detail, is very poor in mainland Britain. Given that good evidence for the manufacture, consumption and ritual use of ale in the Neolithic exists, then it should be possible to find the evidence for sites of grain storage and processing activity. However, often the only surviving evidence for early Neolithic buildings that were constructed of wood are the postholes, pits, hearths and pottery sherds. Such minimal archaeological evidence is difficult to interpret accurately.

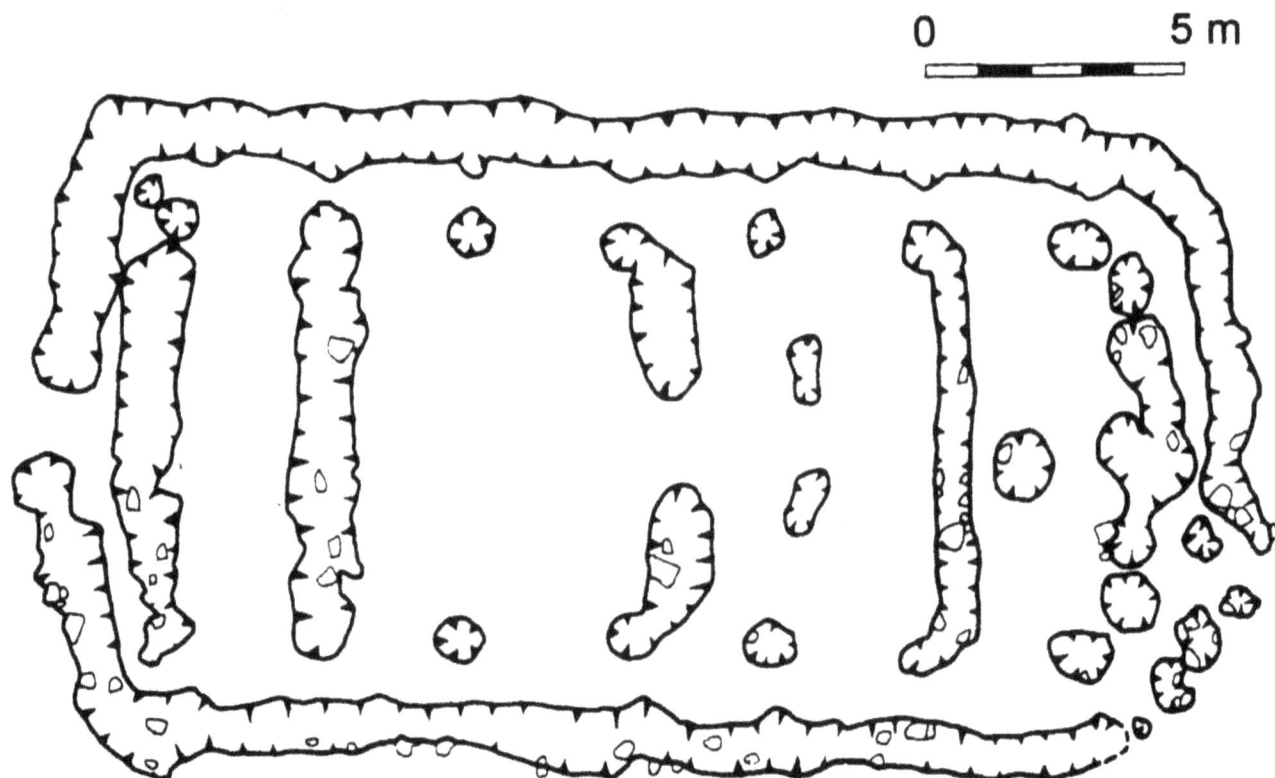

Figure 6.3 Ground plan of Balbridie timber hall, Kincardine, Scotland. It is the largest timber building yet discovered in Britain and Ireland, being approximately 24 metres long and 12 metres wide (Fairweather and Ralston 1993). It was constructed in the early 4ʰ millennium BC and appears to have burnt down. Thousands of carbonised grains were found during excavation. This building was probably used for grain storage and processing.

4. Suitable buildings

A comprehensive catalogue of Neolithic wooden buildings in England, Wales and the Isle of Man has been compiled by Timothy Darvill (1996). He notes traditional beliefs that Neolithic groups either led a nomadic lifestyle or that they built such flimsy dwellings that minimal archaeological evidence remains. Recent discoveries, however, indicate that there was a great diversity of wooden Neolithic buildings, with a wide range of sizes and a variety of purpose, function and design.

To date, 109 certain and probable Neolithic buildings from 64 different sites in England and Wales have been excavated, some rectangular and some circular (Darvill 1996:79). Some have been interpreted as shrines or as buildings where ritual activity took place. Others are interpreted as dwellings. Darvill argues that "a simple binary distinction between structures classifiable as being either ritual or domestic" is no longer viable (ibid:79). Buildings were probably multi functional.

Activities such as malting, mashing and fermentation embody elements of both the mundane and the ritual, as has been extensively argued in earlier chapters. People would have needed both living accommodation and also space for the storage, malting, drying and processing of the grain that they grew. Such a space is today called a Grain Barn or Granary. It could have been controlled, owned or shared by the whole community together. Alternatively, an elite group within the community, as appears to have been the case in the Neolithic Near East (see discussion in Chapter Four), may have had control of the grain store.

Both Graham Clarke (1937) and Stuart Piggott (1954) have referred to the complex dwelling houses, grain barns and granaries of North European cultures. They believed that similar buildings probably also existed in Britain and that, one day, they would be found and excavated (Last 1996:28). Although longhouses like those of the Linearbandkeramik have not yet been discovered, in recent years a number of large rectangular timber buildings, some of which may have functioned as grain barns, have been excavated on mainland Britain. These buildings are represented by only hearth and post hole evidence, nevertheless, some of them may provide evidence for central grain storage and grain processing activities during the early Neolithic in Britain.

Balbridie Timber Hall, Scotland

At Balbridie, Kincardine, Scotland, a very large timber hall was revealed by crop marks and recorded on aerial photographs. It was believed to be of medieval date (figure 6.3). Excavation and radiocarbon dating unexpectedly revealed that it had, in fact, been constructed in the early 4ʰ millennium, c3800 BC (Fairweather & Ralston 1993). Balbridie Hall is approximately 25 metres long by 12 metres wide. It is the largest Neolithic building yet discovered in the British Isles and it has been widely interpreted as a special purpose structure (Thorpe 1996:152). The excavators point out that, as yet, there has been no excavated parallel. This site has yielded the largest assemblage of carbonised grain of Neolithic date yet found in a building in the British Isles.

(b)

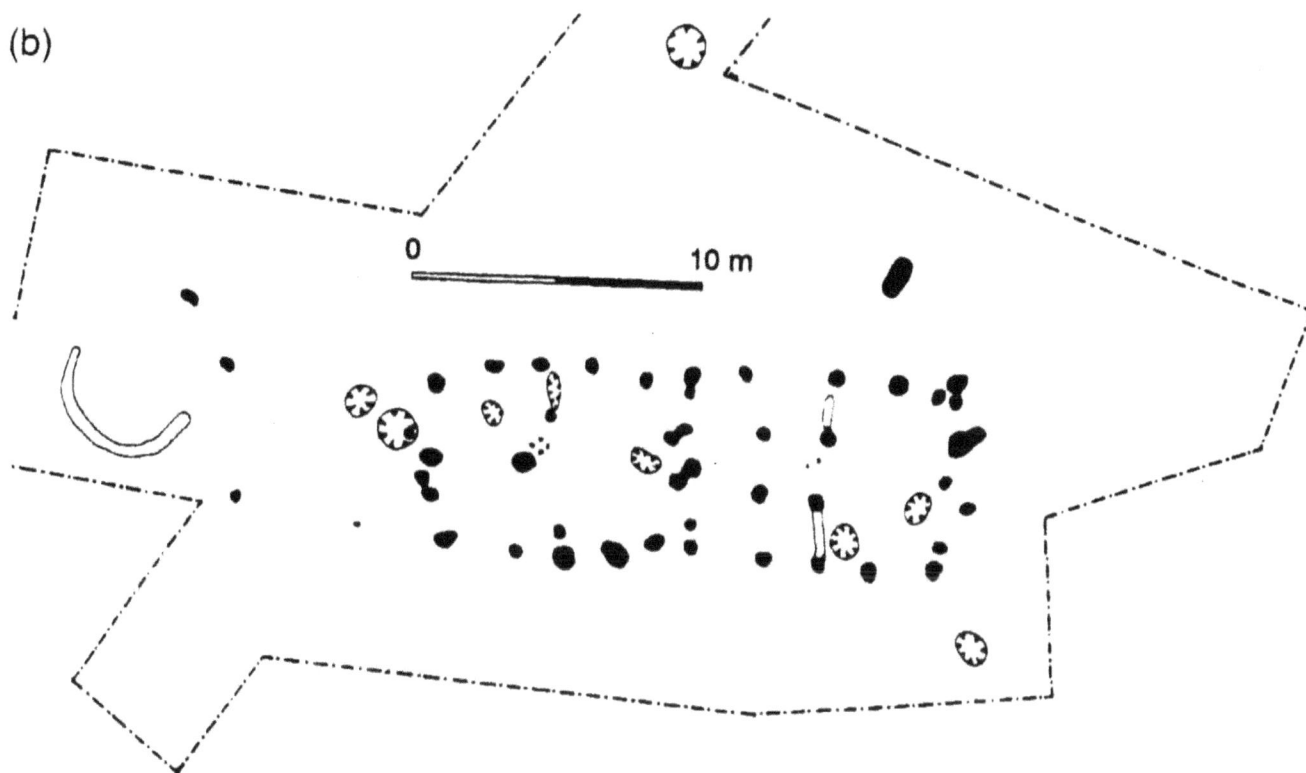

Figure 6.4
Ground plan of one of the two timber structures at Lismore Fields, Buxton, Derbyshire. Radiocarbon dates of c3700 BC were obtained. The structure is approximately 15 metres long and 5 metres wide (Garton 1987). It appears to have been a complex structure, with several hearths. It is difficult to ascertain whether one structure or two adjacent buildings are represented by the postholes. Large quantities of carbonised grain were found here. It may have been malted grain.

About 20,000 cereal grains of Emmer 'bread' wheat and much fewer numbers of naked barley grains were found. The two different grain types appear to have been processed in different areas of the site. The building was being used to store grain and also to process it in different ways. These probably included making bread from the emmer wheat and making malt and ale from the barley grain.

Pottery bowl sherds were also found, representing pottery that is described in the excavation report as being similar in size and style to Unstan Ware pottery. No organic residues survived on any of the pottery sherds for analysis.

Post holes and slot trenches indicate divisions or compartments within the structure. It has been suggested that these are so positioned to prohibit the view inside the building or to 'direct' people in a certain direction. A more practical explanation would be that these screens block sunlight and the direct wind, essential for a malt house. They would also divide the internal space into specific work areas.

Lismore Fields, Buxton, Derbyshire

Perhaps a more suitable candidate for a grain barn of the early Neolithic could be a rectangular timber building, evidence for which has been found at Lismore Fields, near Buxton, Derbyshire (Garton 1987). The building appears to have been divided into four compartments and hearths were set between the lines of postholes that created the divisions (figure 6.4). It was a complex structure with dimensions of approximately 15 metres by 5 metres. Radiocarbon dates of c3700 BC were obtained (Thorpe 1996:152). A wide range of plant fragments was found on the site, including barley chaff, which possibly indicates the malting and mashing of the

grain. Conveniently, a nearby spring would have supplied the necessary water supply.

Looked at practically, this building would provide shelter from the elements and protection from bird or beast attack. The hearths provided the warmth for drying the grain as well as for mashing. The large quantities of carbonised grain found at Balbridie and at Lismore Fields indicate that these buildings were destroyed by fire. This must have been a catastrophic event but it has proved to be providential to archaeologists in the long term, in that the carbonised grain has survived in the archaeological record to give a clue as to one of the probable functions of the building.

Timber buildings, Ireland

In Ireland, over 50 rectangular timber buildings dated to the early Neolithic have been excavated to date and the archaeological evidence involved is complex (Grogan 1996). Some were rectangular and were apparently internally divided into 3 areas, with evidence of more than one hearth. Others were small one-roomed rectangular buildings and yet other buildings were circular, as were most of the structures at Lough Gur, County Limerick, (see Grogan 1996, figures 4.2, 4.3). At Lough Gur there were also large, rectangular timber buildings, apparently not internally divided. Building A had a paved internal area, a central fireplace and an area within the structure perhaps representing the site of a kiln or oven and described as a 'burnt area'. The building may have functioned as a central grain storage and processing area that was used by the whole community. Again, it is a difficult site to interpret from hearth and posthole evidence alone.

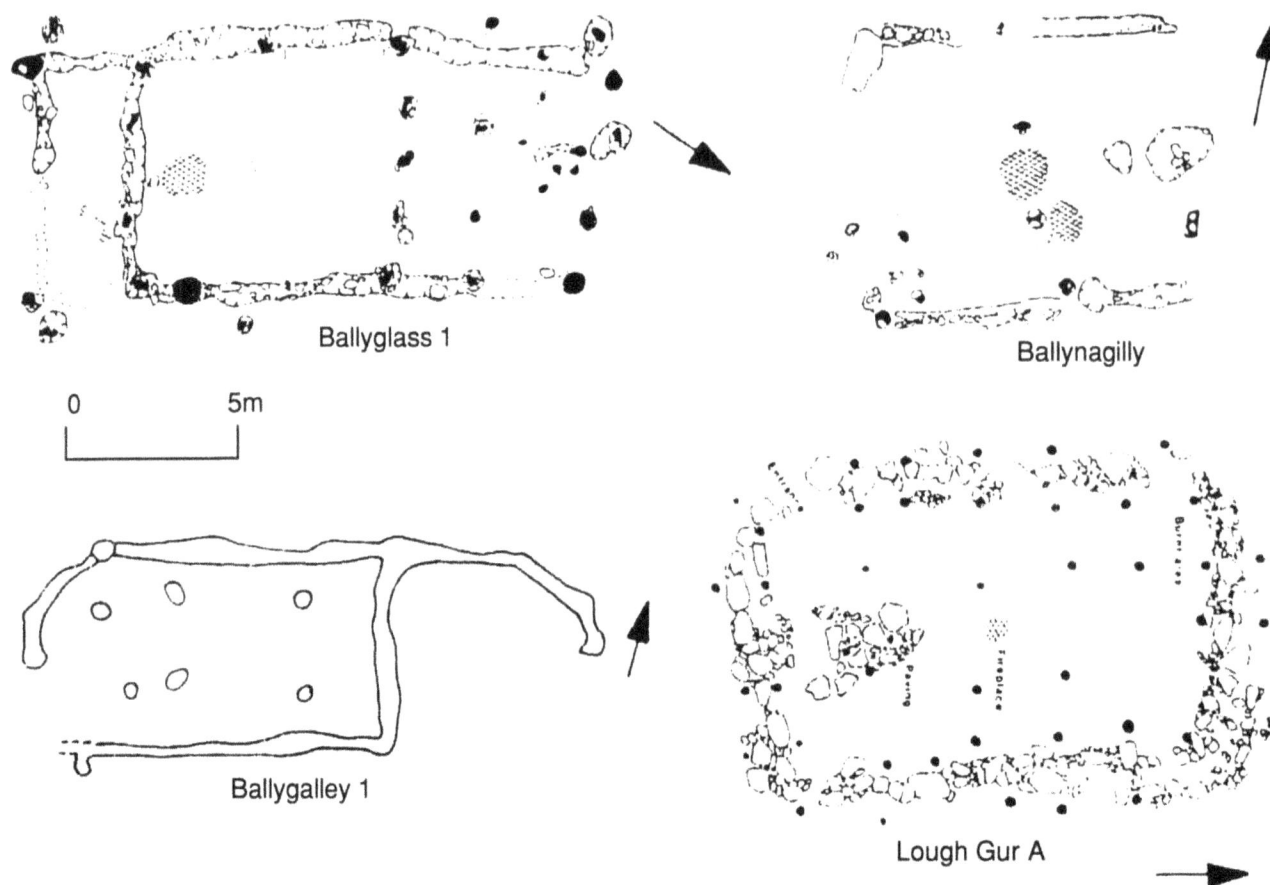

Figure 6.5
Ground plans of some of the Irish timber buildings that might have been used as central grain storage and processing places for early Neolithic communities. There is a wide variety of structural design and it is not easy to be sure of the function of such buildings on post hole evidence alone (after Grogan 1996:45). Arrows indicate North.

Two rectangular houses were found at Balleygalley, County Antrim, (Simpson 1996). House 1 was the better preserved of the two and it is an interesting and unusual building (figure 6.5). Radiocarbon dates of c4226-c3829 BC and c3776-c3386 BC have been obtained from wood charcoal in the bedding trenches and the excavator believes the second date to be the more accurate. Ballygalley has been interpreted as a high status site, possibly a redistribution centre or a storehouse. House 1 contained a number of prestigious items and exotic raw materials, such as rock crystal and over 200 pieces of pitchstone from the Isle of Arran. There was also "a considerable quantity of carbonised grain with what appears to be an unusually high percentage of Einkhorn (*Triticum Monococcum*)" (Simpson 1996:128). No grain processing waste was found from within the house, but spikelets and chaff were recovered from pits nearby, indicating that grain processing was one of the activities of the inhabitants or users of the building.

5. Durrington Walls

Enormous quantities of elaborately decorated Grooved Ware sherds representing some large vessels capable of containing several gallons of liquid were found at the mid 3rd millennium BC site known as Durrington Walls (Wainwright & Longworth 1971). This is a henge enclosure that is almost a third of a mile in diameter (Souden

1997:56,57). It is within the Stonehenge ceremonial landscape and it was a site of feasting, as evidenced by the large numbers of pig bones and hearths that were found there (Albarella & Seargentson 1997). This may be the one of the best examples in the British archaeological record for 'work party' feasting, being situated in the one of the most well known ritual and monumental landscapes of Neolithic Britain.

Durrington Walls henge has only been partially excavated so far and there are more circular and rectangular features within the henge area that are yet to be excavated (figure 6.6). The remains of two timber circles have been excavated and many thousands of sherds from large decorated Grooved Ware vessels were discovered (Wainwright and Longworth 1971). David Souden describes the henge as being "most intimately connected to the River Avon. It lies on a southeast facing slope in a dry valley leading down to the river, and the southeast entrance is only some 60 metres/190 feet from the water's edge" (Souden 1997:56). Access to water is one of the crucial aspects and necessities of brewing and it is very likely that ale was being made and consumed at Durrington Walls.

Outside the henge area itself, the remains of two rectangular timber structures, perhaps barns, have been discovered and excavated according to Timothy Darvill's database (Darvill 1996). Grooved Ware and worked flint were found in association with one post-built, rectangular

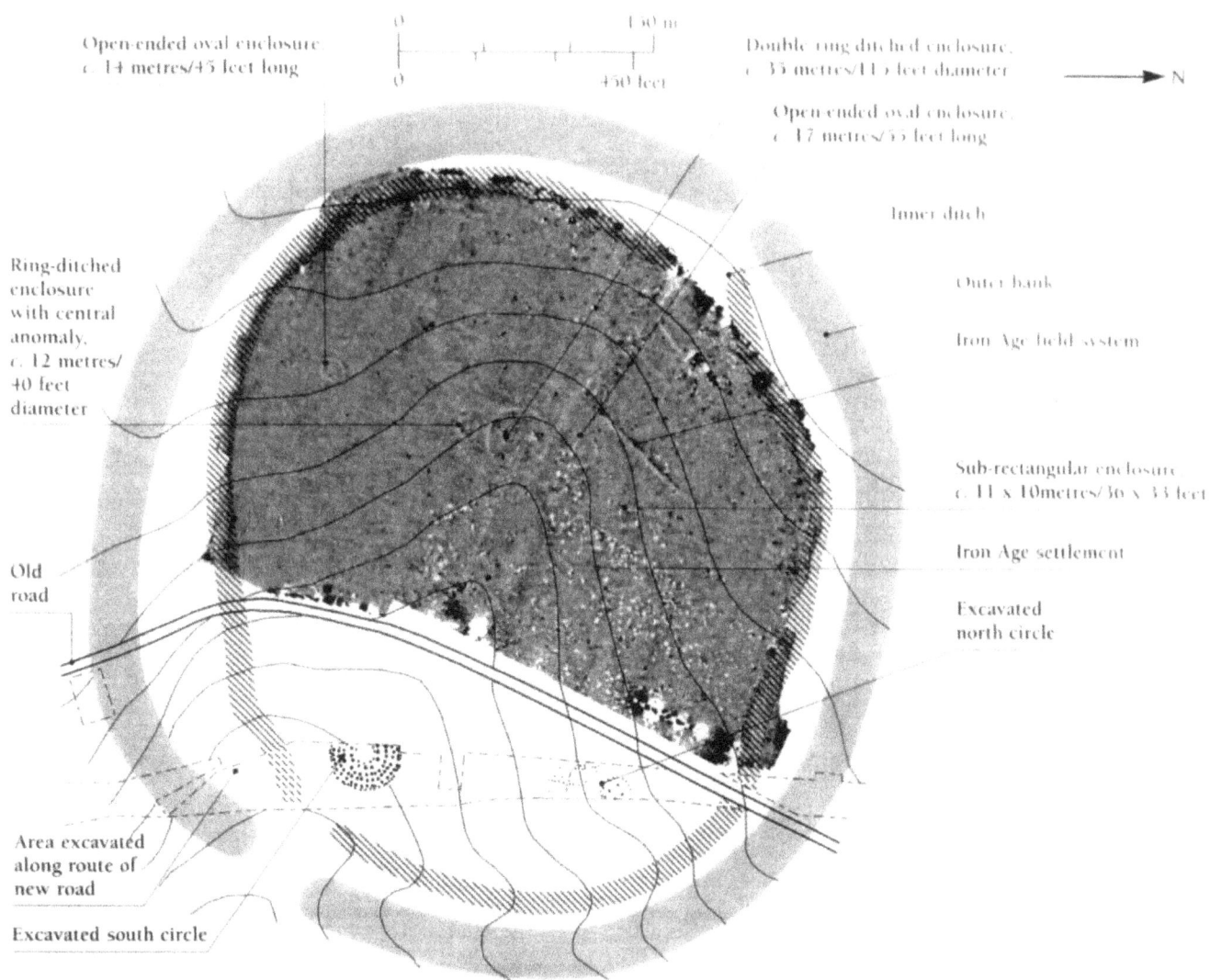

Figure 6.6
Ground plan and aerial photograph of Durrington Walls, showing features excavated by Wainwright in the 1960s and also the details revealed by the 1996 magnetometer survey. The aerial view of the site shows the extent of the enclosure and its proximity to the river, seen at the top left of the photograph (after Souden 1997)

structure, its dimensions being 13 metres by 9 metres, and it is referred to as Durrington 68 (Darvill 1996:107, Pollard 1995). Another rectangular timber building, measuring 18 metres by 10 metres and associated with Grooved Ware, referred to as Totterdown (Structure A), was excavated in 1966 (Wainwright and Longworth 1971). Such buildings may have been used for grain storage and processing.

6. Interpreting Neolithic timber buildings

The accurate interpretation of Neolithic settlement sites and timber buildings on the basis of ground plans and postholes alone is a difficult proposition. A conference organised by the Neolithic Studies Group on Neolithic Houses of North West Europe (Darvill & Thomas 1996) has produced vast amounts of data and information, too detailed and sites too numerous to reproduce and discuss in full here. Much excavation, work and research in the area of Neolithic settlement and the interpretation of buildings on posthole, ceramic, palynological, faunal and other evidence is currently in progress and this is a relatively new area of Neolithic study. Because of the volume of data, not much has yet been analysed with respect to research into Neolithic grain processing techniques and it is hoped to pursue this aspect in the future. Neolithic buildings would probably have fulfilled a variety of needs for the local population, with grain processing only one of several activities.

Julian Thomas (1996) has described timber structures as being "places for meeting and gathering, within which particular activities may have been undertaken in seclusion". He believes them to be a form of 'monumental architecture' and, even though only the postholes of these structures now remain, they were impressive structures in their day. He concludes that "at a number of sites it may be that the 'house' structures form one element in complex sequences of activity which were not purely domestic in character" (Thomas 1996:9-12). Some of these buildings fulfill all the necessary criteria to function as grain storage and grain processing centres for the local community. For early Neolithic groups, perhaps, these buildings were much more than a domestic or a functional building. Important ritual activity took place there too.

The complex sequences of action and activity involved in transforming grain into malt and ale cross the boundaries between the traditional concepts of 'domestic' and 'ritual' activities. The set of specific activities necessary to manufacture malt and ale from the grain is a domestic ritual, requiring skill and knowledge. This has been argued extensively in earlier chapters of this thesis. It was an important part of Neolithic life and may have been a ritualised, even secretive, process. The consumption of the ale at specific occasions, for example at funerals or at rites of passage during life or during celebrations, was both ritually and socially important.

Timber houses and buildings of the Neolithic are the embodiment of complexity. Their interpretation involves a detailed analysis of architectural construction methods as well as the careful consideration of the material culture associated with them including, for example, the artefacts and any organic remains or residues found on site. Neolithic buildings were diverse in form and character and they were far more numerous and widespread than has been commonly believed to date (Darvill 1996:79). As stated earlier, 109 certain Neolithic buildings have been discovered to date from 64 sites in England and Wales dating to the 5th and 4th millennia BC.

Not all of these excavated sites have been fully published and the whole area of interpretation of Neolithic timber buildings is a new and difficult one. Add to this the poor survival of floor and ground surfaces as well as the lack of organic material and any structural remains above ground and the extent of the problem becomes apparent.

For these reasons a detailed study and an in depth analysis of these buildings is impossible within this thesis. There are few sites with favourable conditions, some with "quite appalling" conditions of survival and some with "tantalising" elements (ibid: 81). Post holes and beam slots

with little or no surviving organic material or artefacts are very difficult to interpret. However, within the database compiled by Timothy Darvill (1996:100-111), there are a some timber buildings that it might be possible to interpret as barns or as grain stores on the basis of having ground plans that are similar to the grain barn at Corrigall Farm, Orkney and having artefacts that are connected with grain processing.

At Llandagai, near Bangor, Gwynedd, excavations in 1966-67 (Houlder 1968) revealed remains of a post built structure measuring 13 metres by 6 metres and associated with a nearby henge. Pits and postholes surrounded the building and pottery bowls of the Grimston-Lyles Hill style were associated with the building. A single radiocarbon date of between c4240 BC and c3824 BC determined the date of the structure (ibid:108). Excavations around 1990 at Chigborough Farm, Maldon, Essex revealed the remains of one or perhaps two rectangular timber buildings (Adkins and Adkins 1991). If it were a single structure, considered by Darvill to be more likely, then the dimensions would have been 12 metres by 7 metres (Darvill 1996:103). The ground plan is apparently confusing and unclear.

Grooved Ware and Ronaldsway style pottery sherds were found in association with a rectangular post built structure at Ronaldsway, near Castletown on the Isle of Man (ibid:110). The building was about 7 metres long and 4 metres wide. Artefacts included decorated stone plaques, abundant flint work, animal bone and stone axes. Radiocarbon dates, apparently obtained from residues on pottery, indicate a date of the mid 3rd millennium BC. There was some slight evidence for the building to have been divided internally.

In these few examples cited above there might be some evidence to perhaps interpret that these buildings functioned as Barns, for storing and processing grain or for keeping animals in or for both activities. Suitable internal divisions would have been necessary to keep the beasts away from the grain. The evidence for pottery bowls together with large and small Grooved Ware buckets is probably the best and clearest indication that grain processing was an activity associated with a particular rectangular timber building.

7. The Grooved Ware culture

Grooved Ware pottery is associated with the henges and the timber circles of Mount Pleasant, Woodhenge, Stonehenge and Avebury (Souden 1997:57,114). It has been found in large quantities within a 5 km radius of the Rudston Monolith and associated cursus monuments in Yorkshire (Castledon 1990:86). Andrew Sherratt has suggested that Grooved Ware had "a special quality that set it aside from everyday containers ... this quality lies in the context in which it was used, and this is likely to have been some form of ceremonial meal with sacred connotations, taken at central cult places throughout the length of Britain" (Sherratt 1991:55). If the consumption of the contents of Grooved Ware pottery vessels had connections with sacred, ritual or ceremonial activity, then surely the manufacture of these special contents might also be regarded as a ritually significant activity.

Barley appears to have been an extremely important, perhaps even a sacred crop for people in Britain during the early Neolithic. The cultivation of grain and the manufacture of ale was very probably the task of certain women within the community, as has been argued in previous chapters. Some of these women would have held positions of high status and importance during their lives. They may have been responsible for other sacred and special duties, such as healing people and the provision of medicines, the use of herbs and the control of ritual activities and celebrations.

The people that made and used Grooved Ware pottery were the first grain cultivators and processors in the British Isles. They maintained an active and culturally important hunting lifestyle continuing to exploit the natural resources of the land while they gradually, over the generations, began to adopt what we now recognise as an agricultural or farming lifestyle. These people were the henge builders and the constructors of spectacular monuments of stone and timber.

It has been noted that there is "an absence of grain impressions in Grooved Ware pottery" even though it is held as an established fact that grain cultivation was an important cultural component of the Grooved Ware economy and lifestyle (Jones 1980:62). Martin Jones has suggested that "we should be looking for a behavioural rather than an economic explanation" for this apparent anomaly (ibid: 63). The research presented here is based upon malting and brewing science and valid reconstructions of ancient malting, mashing and fermentation techniques. It would suggest that the behaviour referred to by Jones may have been the ritual processing of barley grain into malt, malt sugars and ale.

Summary and Discussion

This research began as an investigation into the probable methods and techniques of Bronze Age maltsters and brewers in the British Isles and Northern Europe. The original intention was to re-create a Bronze Age ale, based on organic residue evidence that has been discovered in Beaker drinking vessels (Dickson 1978, Barclay et al 1983) and using similar equipment to that available during the Bronze Age.

Debate within academic archaeology about wine, ale, beer and other alcoholic drinks usually tends to concentrate on the social aspects of consumption rather than on the practicalities of manufacture. With this research I aimed to redress the balance and explain the fundamental biochemical reactions and processes that are involved in the malting, mashing and fermentation of grain into ale. These processes remain unchanged across the millennia and allow us to understand something more of past rituals and daily lives.

It has been argued that beer drinking was a Bronze Age phenomenon and that it was part of a 'cult package' that spread across Europe from one group to another (Burgess & Shennan 1976:312). More recent analysis of the origins of alcoholic drinks suggests "the most plausible scenario for the beginnings of alcohol production lies in the domestication of the sugar-rich tree crops of the Mediterranean", such as date, olive, fig, grape and pomegranate (Sherratt 1985:25). Certainly, this is an area where the fermentation of naturally occurring fruit sugars could have been the impetus for the first fermented wines. But beer and ale are products of the grain, a crop that was first gathered and processed by epi Palaeolithic and early Neolithic groups in the Levant and the Near East from the 9th millennium BC onwards.

Grain cultivation and grain processing eventually spread from the Near East and Levant across Europe reaching the British Isles c4000 BC. This research has covered a broad geographical and archaeological range. It has investigated the evidence of the earliest grain processing communities in the Neolithic Levant, Near East, Northern Europe, Orkney, Ireland and Britain. These early agricultural communities have been investigated with one question in mind - was the material culture suitable for the manufacture of malt and ale from the grain? The emphasis throughout this research has been on the practicalities and the specific rituals that are involved in the transformation of grains into malt, malt sugars and ale.

Ian Hodder (1997) has argued that there is a need for archaeologists to question long-held assumptions and 'taken-for-granted' interpretations in archaeology. He has stressed the need to re-interpret the available archaeological evidence holistically, not just looking at isolated aspects of a site but looking instead at the whole cultural and material assemblage. My research has taken this approach.

It has been assumed in most of the archaeological literature that barley, wheat and other cereal grains were a source of carbohydrate in the prehistoric diet and that grain was grown to be processed into only bread, flour, porridge or gruel. It has also been assumed that the main function of quern stones was to grind grain into flour for bread. Querns are just as useful to crush the malted grain prior to mashing.

A search through the index and contents lists of books and articles related to the Neolithic cultures of Europe and Britain shows a significant lack of references to malt, beer, ale, grain processing techniques, brewing or malt sugars. Cereal grains, when discovered in the archaeological record, are often noted as having been 'parched' or 'roasted', the assumption being that the grain has simply been harvested and then dried for optimum storage. If this 'parching' of grain occurred after the grain had begun to germinate then the necessary enzymes to convert the starch into sugars have been released. With a minimum of equipment and resources the malted and dried barley could then very easily be transformed into a sweet malt liquid that can then be fermented into an alcoholic drink, such as beer or ale.

In order to assess the suitability of Neolithic material culture for such grain processing techniques it is essential to understand the processes, methods and techniques that are involved in malting, mashing and fermentation. Because the biochemical laws governing these processes are unchanged across the millennia it is possible to accurately re-create ancient grain processing techniques experimentally, as shown in Chapter One. The biochemistry is complicated but the techniques and methodologies are relatively simple. Brewing is a craft requiring knowledge, skill, practice and experience to successfully transform grain into ale. Pasteur and Tyndall's experiments into Fermentation in the mid 19th century demonstrated the scientific explanation for a biochemical process that had, for millennia, been believed to be a magical and a spontaneous event.

Wild barley and wheat grew naturally in the area known as the Fertile Crescent, that is, the Levant, the Tigris/Euphrates valley and the mountains in northern Syria. Cereal grains were first gathered by Natufian groups in the 9th/8th millennia BC alongside other species of plants, such as lentils and peas. These people were hunters and gatherers. They exploited the natural resources of their environment to the full and this, of course, would have included the gathering of wild grain.

If these wild cereal grains were allowed to grow a little before being ground or crushed with stones, then people would have noticed that there was an obvious visible and practical benefit - the husk of the grain would be broken down and malt flour would be produced naturally. The task of crushing or grinding slightly germinated grain is much easier than crushing ungerminated grain. Invisibly, germination has released enzymes that convert the starch of the grain into malt sugars and produce malt flour. Any gentle heating of the now malted and crushed barley with water would produce a sweet barley mash and malt liquid, so long as the enzymes were not killed in water temperatures that were too hot, that is, above 67 degrees Centigrade.

This saccharification of the barley malt can be seen, smelt and tasted. Knowledge of the existence of enzymes and an understanding of the complex enzymatic reactions are not necessary for this simple process of mashing to be successful. Malting and the subsequent mashing of grains were perhaps among the first grain processing activities in the Fertile Crescent. It is easy to appreciate the wonder and the amazement of these early Neolithic cultures when first introduced to this phenomenon. Here was a food resource that could be processed into sweetness. It was quite unlike other food processing activities that would have been involved with the other locally gathered natural resources, such as the preparation and cooking of peas or lentils. Prior to the discovery of sweet barley mash, the only other source of sugars would have been fruits or honey. The Biblical lands, that is, the area of the Levant, are known as and referred to as 'a Land flowing with Milk and Honey'. Could this description originally have referred to a land where milk was obtained from domesticated animals and sweet malts were processed from the grain, a land where people had learnt to tame the wild animals and to process grain into sweetness?

Malt liquid and barley mash are easy to make, versatile food products and they are very good to eat. They can be mixed with milk to make a delicious and highly nutritious food resource. Malt contains digestible B-Vitamins that would have improved the health of those who began to eat it, although the evidence of the early Natufians' dental caries might suggest an adverse effect on their dental health.

The step from the malting and mashing of barley to that of alcoholic fermentation is not a difficult one to imagine. Wild yeasts would have flourished in a sweet mash or in malt liquids that had been left to stand. Within covered vessels, conditions are perfect for an alcoholic rather than a lactic fermentation. With careful observation, practice and experimentation, the earliest grain processors would have learnt to manage the several stages from grain to ale. These techniques would then have been passed on from one generation to the next.

Because of its unique properties, grain was probably regarded as a special or as a sacred crop in Neolithic times. There were many complex rituals surrounding the cultivation, harvesting and processing of the grain in both prehistoric and historic times. Many of these rituals are still celebrated today, such as the annual Harvest Festival, although now within the context of the Christian religion rather than pagan female deities.

Hilda Ellis Davidson (1998) has produced a detailed and fascinating study of the various goddesses who were worshipped in Northern Europe in prehistoric and historic times. She discusses the important part played by women in ancient and prehistoric cultures. It is the culmination of many years that she has spent studying North European mythology, legends and traditions. She notes that there are very many complex rituals surrounding grain cultivation, such as the preparation of the ground for the sowing of the grain in spring (Davidson 1998:58-68). Some are described in historical texts and some are evidenced by archaeological finds of ards buried in ritual contexts, for example a perfectly preserved ard was discovered in 1994, buried in the ditch of a henge near Dumfries and dated to the early 3rd millennium BC (ibid:60) She describes the ritual uses of the plough, for marking territorial or village boundaries (ibid:64) and for the cutting of ritual furrows to celebrate the beginning of Spring and the new season for cultivation (ibid:59). There were strong associations between the sowing of the seed and the goddess of the grain, with springtime rituals involving human and animal sacrifice taking place in Northern Europe until the 19th Century AD (ibid: 67).

Davidson's work has been referred to occasionally throughout this thesis, but since the emphasis of my study has been on the practicalities of the manufacture of malts and ale,

many of her ideas were not raised or discussed in the earlier Chapters. It is therefore apt to refer to her work here in the context of a discussion of ritual activity, belief systems, magic, barley and the Neolithic. She writes of a rapidly growing interest in the importance of "women as innovators in many fields at a time when small nomadic communities were extending their activities from hunting and gathering to herding and agriculture" (Davidson 1998:8). Women in prehistoric times were largely responsible for the sowing of seed, for the raising of crops, for the processing of grain and for the preparation of a variety of foodstuffs in early agricultural communities. Women also grew and gathered the herbs required for healing and the treatment of injuries, as well as being skilled in midwifery and in the nurturing of children (ibid:154). She concludes that women were the mainstay of the domestic environment, caring for house, home and all within it.

These are very similar ideas to those of Ian Hodder (1990) who has proposed the 'domus' and 'agrios' theories of social development in Neolithic Europe. Davidson's emphasis is on female influence and female power in prehistory. Her book concentrates upon the many and varied "special skills and mysteries of women" one of which was the cultivation and specialised processing of the grain into ale (Davidson 1998:138). Her work is an invaluable study and it should be read by any archaeologist who wishes to better understand the role of the female in prehistory, as well as the ancient belief systems and rituals of the past. It provides an insight into many aspects of domestic, spiritual and ritual life in prehistory.

Early Neolithic communities in the Levant and the Near East would have learned the necessary methods and techniques of grain processing through repeated trial and error. This knowledge then spread through the complex and far-reaching trade and exchange networks that made use of land, sea and river routes into Europe. The practice of and the ideas behind grain cultivation and processing spread rapidly into northern Europe. The similarities in lifestyles and material culture of the inhabitants of the Bulgarian settlement tells of the 6th/5th millennia BC and those of 6th millennium BC in Anatolia have been noted in Chapter Four of this thesis. Many more archaeological examples could have been selected for a similar comparison. For example, Alisdair Whittle (1996) describes the Vinca culture of the 6th and 5th millennia BC and the elaborate and complex late Neolithic buildings of the Hungarian plain. These were cultures that also possessed the basic requirements for making malt and brewing ale.

The coastal groups and communities of northern Europe maintained a hunting, gathering and fishing lifestyle for almost a millennium longer than the agricultural communities in Central Europe, who lived in settlements along the river valleys. However, there is some evidence of organic residues on Ertebølle pottery vessels that have been interpreted as being the remnants of fermented grain and blood (Tilley 1996:25). This suggests the probable trade and exchange of grain between agricultural and non-agricultural communities of the 5th and 4th Millennia BC. It also suggests interesting and as yet unknown ritual behaviour in both the manufacture and in the consumption of this alcoholic drink. Eventually, the Northern European Mesolithic groups began to cultivate their own grain, some time in the early 3rd millennium BC.

Many different theories have been put forward concerning the reason for the acceptance of grain cultivation within these groups and some of these have been discussed in Chapter Four. It seems likely that Mesolithic groups were interested in the products of the grain, that is the sweet malts and the ale, rather than a change of lifestyle to that of farming.

'Farming' is a cultural concept and use of the word creates an image of an organised and regulated farmstead as we are accustomed to seeing today or in historical times. Farming is a way of life that has evolved and developed over the years as a result of people's desire to grow and to process grain and other crops and to keep domesticated animals. To refer to these early cultivators and processors of grain as 'farmers' does not really seem to be appropriate.

One of the most striking aspects of the European and British Neolithic was the construction of huge communal monuments, standing stones and finely constructed tombs. Ritual behaviour and activity is one of the most discussed and well known aspects of the Neolithic. The people who made and used Grooved Ware during the 4th and 3rd millennia BC were the earliest grain cultivators and processors in Britain. They continued to exploit the natural resources of the seas, the rivers and the woodland whilst they began to cultivate grain and to manufacture malts and ale, as argued in Chapters Five and Six of this thesis. They also constructed impressive and lasting monuments, such as the two stone circles on Orkney, numerous timber and stone circles throughout the mainland, elaborate tombs and burial chambers as well as standing stones, cursus monuments and henges. My research indicates that there was a powerful female element to this culture that was closely related to ritual activities and to the cultivation and processing of barley.

Organic residues containing potentially dangerous psychoactive substances, such as the crushed Henbane seeds that were discovered on Grooved Ware sherds at Balfarg (Barclay et al 1993), indicate that mind-altering alcoholic brews were sometimes made for ritual occasions. It is impossible to know whether the purpose of this brew was for shamanic and magical practices or as a poisonous drink for use in a ritual funerary context. Its potential use as an 'external medicine' as noted by Culpepper, perhaps for toothache, cannot be ignored.

This research, which began as being a relatively straightforward examination of the likely brewing methods of British Bronze Age people has revealed many fascinating and previously unconsidered aspects of Neolithic life. There is much further work to be done in this area. The role of women in the Neolithic needs to be re-evaluated, for example, what was their role in healing, medicine and in ritual activity and what was their knowledge and use of herbs? Archaeobotanical study and analysis can be very useful in answering these questions. A serious plea has to be made to all archaeologists to retain and to analyse the organic residues on pottery sherds rather than destroy such important evidence by routinely scrubbing the pottery.

One aspect of the Neolithic that has been unexpectedly illuminated by this research is the importance of malt in prehistory. Although the manufacture and the consumption of ale and of other alcoholic drinks is seen as being important ritually, socially and economically, the manufacture of the malt may be just as, if not more, important. Mixed with milk or eaten as a product in its own right, malt would have been a nutritious addition to Neolithic diet and appealed to young and old alike. In the public demonstrations and tastings of the barley mash that I have undertaken as part of this research the overwhelming response has been positive. People have expressed a liking for the sweet mash and return for a second tasting.

Conclusions

This research has established an assemblage and a material culture pattern for brewing activity in prehistory. Suitable buildings are required for grain storage and for malting and otherwise processing the grain. A malting floor can be made of beaten earth or clay and needs to be kept smooth and in good repair. Hearths, ovens or kilns are useful for drying the malt and as a heat source for mashing and fermentation. Suitable vessels for mashing, fermentation, storage and consumption must be made and access to running water and/or drains is essential. Such conditions and material culture are good indicators of malting, mashing and brewing activity.

Women were the very first grain cultivators and processors in the Near East, the Levant, Europe and the British Isles. Grain was a special crop because of its unique ability to produce sugars. Women, with their understanding of grain cultivation and processing rituals and their knowledge of the use of wild plants and herbs for both culinary and medicinal uses, held positions of status and significance in Neolithic society.

Brewing uses few ingredients, only requiring malted grain, herbal preservatives, water and yeast. These ingredients may survive in the archaeological record in a number of ways. Accidents in drying the malted grain, as happened at Eberdingen-Hochdorf (figure 1.5) can occur. Residues or sediments of the brewing process may occasionally survive in unusual contexts, such as in the sealed Bronze Age cist graves at North Mains and Ashgrove (figures 1.16, 1.17). Residues of barley without any other plant remains indicate the residues that result from washing the sugars from the mashed barley or 'sparging the wort' (figure 1.7). Those barley residues that contain pollen or macro plant remains indicate the addition of herbs during the boil prior to fermentation.

The ease with which the barley malt and mash can be made convinces me that the manufacture of these products was a main interest and concern in the collection and cultivation of grains by Epi-Palaeolithic and Natufian cultures. The production and manufacture of this liquid product would have created a need for vessels and containers that were suitable for the storage and processing of the product, hence the bitumen lined baskets and experiments with White Ware and ceramics. The spread of grain cultivation and processing from the Levant across Europe and into the British Isles was accompanied by a developing ceramics technology and the domestication of animals. The animals would have eaten the 'spent grain' with as much relish as people, adults and children ate the sweet malt products and drank the ale.

BIBLIOGRAPHY

Adkins, K and Adkins, P C 1991 'A Neolithic settlement on the north bank of the river Blackwater.' *Colchester Archaeological Group Annual Bulletin* 34, 15-28.

Albarella, U and Sarjeantson, D 2002 'A passion for pork: meat consumption at the British late Neolithic site of Durrington Walls.' In Miracle, P (ed) *'Consuming Passions and Patterns of Consumption.'* The McDonald Institute for Archaeological Research. Publication from a conference held at the Institute, University of Cambridge, September, 1997.

Appleby, A 2000 'Neolithic pottery experiments.' Part of Chapter 16 in Ritchie, A (ed) *Neolithic Orkney in its European Context.* McDonald Institute Monographs, Oxbow Books. Conference, Kirkwall, Orkney, September 1998.

Armit, I and Finlayson, B 1992 'Hunter-gatherers Transformed: The Transition to Agriculture in Northern and Western Europe.' *Antiquity* 66, 664-676.

Ashmore, P 1996 *Neolithic and Bronze Age Scotland.* Historic Scotland, Batsford.

Bakels, C 2000 'The Neolithization of the Netherlands: Two ways, one result.' In Fairbairn, A (ed) *Plants in the Neolithic of Britain and beyond.* Oxbow Books. Spring Meeting, Neolithic Studies Group, London, March 1998.

Bakker, J 1979 *The TRB West Group: Studies in the Chronology and Geography of the makers of Hunebeds and Tiefstich pottery.* Universiteit van Amsterdam.

Barclay, G J et al 1983 'Sites of the 3rd millennium BC to the 1st millennium AD at North Mains, Strathallan, Perthshire.' *Proceedings of the Society of Antiquaries of Scotland* Vol 113, 122-282.

Barclay, G J and Russell-White, C J (eds) 1993 'Excavations in the ceremonial complex at Balfarg/Balbirnie, Glenrothes, Fife.' *Proceedings of the Society of Antiquaries of Scotland* Vol 123, 43-210.

Barclay, G 1993 *Balfarg: The prehistoric ceremonial complex at Glenrothes, Fife.* Fife Regional Council.

Barclay, G 1996 'Neolithic buildings in Scotland.' In Darvill, T and Thomas, J (eds) *Neolithic Houses in Northwest Europe and Beyond.* Neolithic Studies Group Seminar Papers I, Oxbow Monograph 57, 61-77.

Bailey, D W 1996 'The life, times and works of House 59, Tell Ovchorovo, Bulgaria.' In Darvill T and Thomas J (eds) *Neolithic Houses in Northwest Europe and Beyond.* Neolithic Studies Group Seminar Papers 1 Oxbow Monograph 57, 143-157.

Belfer-Cohen, A 1995 'Rethinking social stratification in the Natufian culture: the evidence from burials.' In Campbell S and Green A (eds) *The Archaeology of Death in the Near East.* Oxbow Books, Oxford.

Bennett, J 1989 *Sisters and Workers in the Middle Ages.* University of Chicago Press.

Bennett, J 1996 *Ale, Beer and Brewsters in England: Women's work in a changing world, 1300.* Oxford University Press.

Bewley, R 1994 *The English Heritage Book of Prehistoric Settlements.* Batsford, London.

Berglund, B and Kolstrup, E 1991 'The Romele Area: Vegetation and Landscape through Time.' In Berglund B (ed) *The cultural landscape during 6000 years in southern Sweden – The Ystad Project,* Lund: Ecological Bulletin 41, 65-68. (referred to in Thorpe, I 1996).

Biel, von J 1996 *Experiment Hochdorf: Keltische Handwerkskunst Wiederbeleb herausgegeben.* Stuttgart, Wais and Partner.

Blankholm, H 1987 'Late Mesolithic Hunter Gatherers and the Transition to Farming in Scandinavia.' In Rowley-Conwy, Zvelebil, Blankholm (eds) *Mesolithic Northwest Europe, recent trends* Dept Archaeology and Prehistory, Sheffield University.

Bogucki, P 1988 *Forest Farmers and Stockherders.* Cambridge University Press.

Bogucki, P 1997 'The Neolithic Mosaic on the North European Plain.' Paper, Society for American Archaeology, Atlanta, Georgia, 1989. www.princeton.edu/~bogucki/mosaic

Bond, J et al 1995 'Stove Bay: a new Orcadian Grooved Ware settlement.' *Scottish Archaeological Review* Vols 9, 10, 125-131.

Braidwood, R 1953 'Did man once live by bread alone?' *American Anthropologist* 55, 515-526.

Burgess, C and Shennan, S 1976 'The Beaker Phenomenon: some suggestions.' In Burgess, C & Miket R (eds) *Settlement and Economy in the 3rd millennium BC.* British Archaeological Reports 33, 309-327.

Byrd, B F and Moynahan, C M 1995 'Death, Mortuary Ritual and Natufian Social Structure.' *Journal of Anthropological Archaeology* Vol 14, No 3, 251-288.

Castledon, R 1990 *The Stonehenge People: life in Neolithic Britain 4700-2000 BC.* Routledge.

Chapman, J 1991 'The creation of Social Arenas in the Neolithic and Copper Age of SE Europe: The case for Varna.' In Garwood P, Jennings D, Skeates R and Toms J (eds) *Sacred and Profane, Proceedings of a Conference on Archaeology, Ritual and Religion.* Oxford University Committee for Archaeology Monograph No 32.

Bibliography

Childe, V G and Paterson, J W 1929 'Provisional Report on Excavations at Skara Brae and on finds from the 1927 and 1928 campaigns.' *Proceedings of the Society of Antiquaries of Scotland Vol* LXIII, 225-279.

Childe, V G 1930 'Operations at Skara Brae during 1929.' *Proceedings of the Society of Antiquaries of Scotland* Vol LXIV, 158-191.

Childe, V G 1930 'Final Report on Operations at Skara Brae.' *Proceedings of the Society of Antiquaries of Scotland* Vol LXV, 27-77.

Childe, V.G. 1931 'The continental affinities of British Neolithic Pottery.' *Archaeological Journal* Vol 88, 37- 67.

Childe, V G 1931 *Skara Brae; a Pictish Village in Orkney.* London.

Childe V G & Grant W G 1938 'A Stone Age settlement at the Braes of Rinyo, Rousay, Orkney (First Report).' *Proceedings of the Society of Antiquaries of Scotland* Vol LXIII, 6-39.

Childe, V G & Grant, W G 1946-8 'A Stone Age settlement at the Braes of Rinyo, Rousay, Orkney (Second Report).' *Proceedings of the Society of Antiquaries of Scotland* LXXXI, 16-42.

Clarke D L 1970 *Beaker Pottery in Great Britain and Ireland.* Cambridge University Press.

Clark, J G D 1977 *World Prehistory: in new perspective.* (3rd edition) Cambridge University Press.

Clarke, D V 1976a *The Neolithic village at Skara Brae, Orkney, Excavations 1972-3: An interim report.* HMSO Edinburgh.

Clarke, D V 1976b 'Excavations at Skara Brae: A summary account.' In Burgess C and Miket, R (eds) *Settlement & Economy in the Third and Second Millennia BC* British Archaeological Reports 33, 233-250.

Clarke, D V, Cowie, T G and Foxon A 1985 *Symbols of Power at the time of Stonehenge.* HMSO National Museum of Antiquities of Scotland, Edinburgh.

Clarke D V and McGuire, P 1989 *Skara Brae: Northern Europe's best-preserved prehistoric village.* Historic Scotland.

Clark, G 1989 *Economic Prehistory: Papers on Archaeology.* Cambridge University Press.

Collon, D 1995 *Ancient Near Eastern Art.* British Museum Press, London.

Comey, M 1996 *The Archaeology of Ale in Early Christian Ireland AD 400-1200.* Unpublished Undergraduate Thesis, University College London, Institute of Archaeology.

Conant, J B (ed) 1952 *Pasteur's Study of Fermentation.* Harvard Case Histories in Experimental Science, Harvard University Press.

Contenson, de H 1971 'Tell Ramad, a village of Syria, 7th and 6th millennia BC.' *Archaeology* Vol 24, 278-285.

Cooney, G 1999 "A boom in Neolithic houses." *Archaeology Ireland* Spring Issue, 13-16.

Corran, H S 1975 *A History of Brewing.* David & Charles.

Coudart, A 1991 'Social structure and relationships in prehistoric small-scale societies: the Bandkeramik groups in Neolithic Europe.' In Gregg S (ed) *Between Bands and States.* Carbondale: Southern Illinois University Press 295-420.

Crane, E 1983 *The Archaeology of Bees* Duckworth & Co.

Crawford, H 1981 'Some fire installations from Abu Salabikh, Iraq.' *Paleorient* Vol 7/2 105-114.

Culpepper, N 1616-1654 *Culpepper's Complete Herbal,* Foulsham Edition.

Darvill, T 1990 *Prehistoric Britain.* Batsford.

Darvill, T & Thomas, J (eds) 1996 *Neolithic Houses in Northwest Europe and Beyond.* Neolithic Studies Group Seminar Papers I, Oxbow Monograph 57.

Davidson, H E 1988 *Myths and Symbols in Pagan Europe: Early Scandinavian and Celtic Europe.* University of Manchester Press.

Davidson, H E 1998 *Roles of the Northern Goddess.* Routledge.

Devereux, P 1997 *The Long Trip: The Prehistory of Psychedelia.* Penguin.

Dickson, J 1978 'Bronze Age Mead.' *Antiquity* 52, 108-112.

Dietler, M 1989 'Driven by Drink: The role of drinking in the political economy and the case of Early Iron Age France.' *Journal of Anthropological Archaeology* 9, 352-406.

Dietler, M 1996 'Feasts and commensal politics in the political economy: Food, Power and Status in Prehistoric Europe.' in Weissner, P and Schiefenhovel, W (eds) *Food and the status quest: an interdisciplinary perspective.* Berghan Books.

Dineley, M 1996 'Finding Magic in Stone Age Real Ale.' *British Archaeology* November, No 19, 6.

Dineley, M 1997 'Beer brewing formed part of Neolithic ceremonies. *British Archaeology (News)* Sept, No 27, 4.

Dineley, M and Dineley, G 2000 'From Grain to Ale: Skara Brae, a case study.' Part of Chapter 16 in Ritchie, A (ed) *Neolithic Orkney in its European Context.* McDonald Institute Monographs, Oxbow Books.

Dineley, M and Dineley, G 2000 'Neolithic ale: Barley as a source of malt sugars for fermentation.' In Fairbairn, A (ed) *Plants in Neolithic Britain and beyond.* Neolithic Studies Group Seminar Papers 5, 137-155, Oxbow Books. Paper presented at the Spring Meeting, Neolithic Studies Group, Royal Academy, London, March, 1998.

Earwood, C 1993 *Domestic Wooden Artefacts in Britain and Ireland from Neolithic to Viking Times.* University of Exeter Press.

Ebbeson K 1978a 'Stenalderlerkar med ansigt.' *Kuml* 99-115. (referred to in Tilley, C 1996).

Ehrenburg, M 1982 *Women in Prehistory.* British Museum Publications.

Fairweather, A D and Ralston, I B M 1993 'The Neolithic timber hall at Balbridie, Grampian Region, Scotland: A preliminary note on dating and plant macrofossils.' *Antiquity* 67 313-323.

Fischer, A 1981 'Danubian shaft hole axes.' *Journal of Danish Archaeology* Vol 1, 7-13.

Flett, H Curator Corrigall Farm Museum, Harray, Orkney. Conversations, Summer 1998.

Fraser, D 1983 *Land and Society in Neolithic Orkney.* British Archaeological Reports 117, BA R Publishing.

Garton, D 1987 'Buxton' *Current Archaeology* 9.8 (No 103), 250-253.

Garwood P, Jennings D, Skeates R and Toms J (eds) 1991 *Sacred and Profane, Proceedings of a Conference on Archaeology, Ritual and Religion.* Oxford University Committee for Archaeology Monograph No 32.

Gayre G R 1948 *Wassail! : in Mazers of Mead. An account of Mead, Metheglin, Sack and other Ancient Liquors and of the Mazer cups out of which they were drunk, with some comment on the Drinking Customs of our forbears.* London. Philimore and Co. Ltd.

Genders, R 1971 *The Scented Wild Flowers of Britain.* Collins.

Gibson, A 1998 *Stonehenge and Timber Circles.* Tempus Publishing Ltd.

Gilmour, J, and Walters, M 1969 *Wild Flowers.* (Fourth Edition) Collins.

Gimbutas, M 1982 *The goddesses and gods of Old Europe: 6500-3500 BC: Myths and cult images.* Thames and Hudson.

Goodman, J Lovejoy, P E and Sherratt, A 1995 *Consuming Habits: Drugs in History and Anthropology.* Routledge.

Grogan, E 1996 'Neolithic Houses in Ireland' In Darvill T and Thomas J *Neolithic Houses in northwest Europe and Beyond* Neolithic Studies Group Seminar Papers 1, Oxbow Monograph 57, 41-61.

Grygiel, R 1980 Jama ze spalona pszenica kultury pucharow lej kowatych z Opatawic, woj. Wloclawskie. *Prace i Materialy Muzeum Archeologicznego i Ethnograficznego.* 31, 43-334. (referred to in Bogucki 1988)

Haggerty, A 1991 'Machrie Moor, Arran: recent excavations at two stone circles.' *Proceedings of the Society of Antiquaries of Scotland* Vol 121, 51-94.

Halstead, P 1993 'Spondylus shell ornaments from late Neolithic Dimini, Greece: specialised manufacture or unequal accumulation?' *Antiquity* 67, 603-9.

Harbison, P 1994 *Pre-Christian Ireland.* Thames and Hudson.

Harlan, J 1967 'A wild wheat harvest in Turkey.' *Archaeology* XX, 197-201.

Hawkes, J. 1934 'Aspects of the Neolithic and Chalcolithic Periods of Western Europe.' *Antiquity* Vol VIII, 24-42.

Hazzledene, Piggott, S et al 1936 'Archaeology of the submerged land surface of the Essex coast. *Proceedings of the Prehistoric Society* Vol 11 (Part II), 178-211.

Hayden, B 1990 'Nimrods, Piscators, Pluckers and Planters: The Emergence of Food Production.' *Journal of Anthropological Archaeology* 9, 31-69.

Hayden, B 1996 'Feasting in prehistoric and traditional societies. In Weissner, P and Schiefenhovel, W (eds) *Food and the status quest: an interdisciplinary perspective.* Berghan Books.

Hodder, I 1990 *The Domestication of Europe.* Blackwell.

Hodder, I 1997 'Always momentary, fluid and flexible: towards a reflexive excavation methodology.' *Antiquity* 71, 691-700.

Houlder, C 1968 'The henge monuments of Llandegai. *Antiquity*, 42, 216-221.

Hulme, P D and Sherriff, J 1985 "Pollen analysis of a radiocarbon dated core from North Mains, Strathallan." *Proceedings of the Society of Antiquaries of Scotland* 115, 105-13

Hunter J and McSween A 1991 'A sequence for the Orcadian Neolithic' *Antiquity 65* Vol 65, 911-914.

Bibliography

Irving, A 1998 *Approaches to Style in Near Eastern Ceramics.* Unpublished M.Phil Thesis, Department of Art History and Archaeology, University of Manchester.

Isbister, A 2000 'Burnished Haematite and pigment production.' part of Chapter 16 in Ritchie, A. *Neolithic Orkney in its European Context.* McDonald Institute Monographs, Oxbow Books. Conference Kirkwall, Orkney, 1998.

James, P and Thorpe, N 1995 *Ancient Inventions.* Ballantine Books, London.

Joffe, A H 1998 'Alcohol and Social Complexity in Ancient Western Asia.' *Current Anthropology* Vol 9, No3, 297- 322.

Jones, A 1999 'The World on a Plate: ceramics, food technology and cosmology in Neolithic Orkney.' *World Archaeology* 31 (1), 55-78.

Jones, A 2002 *Archaeological Theory and Scientific Practice.* Cambridge University Press.

Jones, M 1980 'Carbonised Grains from Grooved Ware contexts.' *Proceedings of the Prehistoric Society* 46, 61-63.

Kafafi, Z A 1986 'White objects from Ain Ghazal, Near Amman.' BASOR 261.

Katz, S and Voigt, M 1986 'Bread and Beer: The early use of cereals in the human diet.' *Expedition* Vol 25/2, 23-34.

Katz, S and Maytag, F 1991 'Brewing an Ancient Beer.' *Archaeology*, Vol 44 No 4, 24-33.

Kinnes, I 1985 'Circumstance not Context: The Neolithic of Scotland as seen from the outside.' *Proceedings of the Society of Antiquaries of Scotland* 115, 15-57.

Krausse, D 1996 *Hochdorf III* Kommisionsverlag, Konrad Theiss Verlag, Stuttgart.

Kretschmer, von H 1996 'Brauen fruher und heute.' In Biel 1996 *Experiment Hochdorf: Keltische Handwerkskunst Wiederbelebt.* Stuttgart 76-82.

Last, J 1996 'Neolithic houses, a central European perspective.' In Darvill, T and Thomas, J (eds) *Neolithic Houses in Northwest Europe and Beyond.* Neolithic Studies Group Seminar Papers I, Oxbow Monograph 57, 27-41.

Lehane, B 1977 *The Power of Plants.* John Murray, London.

Line, D 1985 *The Big Book of Brewing.* (14th Printing) Argus Books Ltd, G W Kent Inc, USA.

Long, D J, Milburn, P, Bunting, J M, Tipping, R and Holden, T 1999 'Black Henbane in the Scottish Neolithic: A Re-evaluation of Palynological findings from Grooved Ware pottery at Balfarg Riding School and Henge, Fife.' *Journal of Archaeological Science* Vol 26, 45-52.

Lucas, F 1962 (4th edition) *Ancient Egyptian materials and industry.* W & J Mackay & Co.

Mabey, R 1996 *Flora Britannica.* Sinclair Stevenson.

McGee, H 1984 *On Food and Cooking.* Allen & Unwin.

MacKie, E 1997 'Maeshowe and the winter solstice: ceremonial aspects of the Orkney Grooved Ware Culture.' *Antiquity* 71, 338-59.

MacSween, A 1992 'Orcadian Grooved Ware.' In Sharples N and Sheridan A *Vessels for the Ancestors.* University of Edinburgh Press.

Maisels, C K 1990 *The Emergence of Civilisation: from hunting and gathering to agriculture, cities and the state in the Near East.* Routledge, London.

Maisels, C K 1993 *The Near East: archaeology in the cradle of civilisation.* Routledge.

Mellaart, J 1967 *Catal Huyuk.* Thames & Hudson.

Mellaart, J 1975 *The Neolithic of the Near East.* Thames and Hudson.

Metzner, R 1994 *'The Well of Remembrance' Rediscovering the Earth Wisdom Myths of Northern Europe.* Shambhala Publications.

Michel, R H and McGovern, P 1992 'Chemical Evidence for Ancient Beer. Correspondence in *Nature* Vol 360, 24.

Molleson, T 1994 'The Eloquent Bones of Abu Hureyra. *Scientific American* Vol 271, 60-66.

Moore, J 1981 'The effects of information networks in hunter gatherer societies.' In *Hunter Gatherer Foraging Strategies* Winterlader B and Smith E A 194-217 (referred to in Bogucki 1988)

Pasteur, L 1879 (tr. Faulkener, F and Robb C) *Studies on Fermentation. The diseases of beer: their causes and the means of preventing them.* MacMillan & Co.

Petrie, F 1867 'Notice of ruins of ancient dwellings at Skara, Bay of Skaill, in the parish of Sandwick, Orkney, recently excavated. *Proceedings of the Society of Antiquaries of Scotland* Vol VII, 201-218.

Piggott S. 1931 'The Neolithic pottery of the British Isles.' *Archaeological Journal* 88, 67-160

Piggott, S. 1954 (reprinted 1970) *The Neolithic Cultures of the British Isles.* Cambridge University Press.
Pliny *Natural Histories.* Book XIV Chapter 29.

Pollard, J 1997 *Neolithic Britain.* Shire Archaeology.

Ratsch, C 1994 'The Mead of Inspiration and Magical Plants

of the Ancient Germans.' In Metzner, R *The Well of Remembrance: rediscovering the Earth Wisdom Myths of Northern Europe*. Shambhala Publications.

Renfrew, C 1985 *The Prehistory of Orkney*. Edinburgh University Press.

Renfrew, J 1985 *Food and Cooking in Prehistoric Britain: History and Recipes*. English Heritage.

Richards, C 1991a 'Skara Brae: revisiting a Neolithic village in Orkney.' In Hanson and Slater (eds) *Scottish Archaeology: New Perceptions*. Aberdeen University Press.

Richards, C 1991b 'The Late Neolithic house in Orkney. In Sampson, R E (ed) *The Social Archaeology of Houses*. Edinburgh University Press, 111-124.

Richards, C 1992 'Barnhouse and Maeshowe.' *Current Archaeology* 131, 444 - 448.

Richards, C 1996 'Monuments as landscape: creating the centre of the world in late Neolithic Orkney.' *World Archaeology: Sacred Geography* Vol 28 (2), 190-208.

Richards, M 1996 'Early farmers with no taste for grain.' *British Archaeology* March No 12, 6.

Ritchie, A 1983 'Excavation of a Neolithic farmstead at Knap of Howar, Papa Westray, Orkney.' *Proceedings of the Society of Antiquaries of Scotland 113*, 40-121.

Ritchie, A 1985 'The First Settlers.' In Renfrew, C *The Prehistory of Orkney*. Edinburgh University Press.

Ritchie, J N G 1976 'The Stones of Stenness, Orkney.' *Proceedings of the Society of Antiquaries of Scotland*. Vol 107, 1-60.

Rose, A 1977 *Alcoholic Beverages*. Academic Press, London.

Rose, F 1991 *Wild Flower Key*. Warne.

Rollefson, G 1983 'Ritual and ceremony at Neolithic Ain Ghazal.' *Paleorient* Vol 9/2, 29-38.

Rollefson, G 1986 'Neolithic Ain Ghazal, Jordan: Ritual and Ceremony II.' *Paleorient* Vol 12/1, 45-52.

Rollefson, G and Simmons, A 1987 'The Life and Death of Ain Ghazal.' *Archaeology* Vol 40, 38-45.

Rollefson, G 1989 'The aceramic Neolithic of the Southern Levant: the view from Ain Ghazal.' *Paleorient* Vol 15/1 135-40.

Rowley-Conwy, P 1985 'The Origins of Agriculture in Denmark: a review of some theories.' *Journal of Danish Archaeology* Vol 4, 188-196.

Samuel, D 1995 'Rediscovering Ancient Egyptian Beer.' *Brewer's Guardian*, Vol 124, No 12.

Samuel, D 1996 'Archaeology of Ancient Egyptian Beer.' *Journal of the American Society of Brewing Chemists* 54 (1), 3-12.

S.E.A. Scottish Ethnographic Archive. This provides information from press cuttings giving details of brewing techniques in early historical times and in recent history.

Sharples, N and Sheridan, A 1992 *Vessels for the Ancestors*. Edinburgh University Press

Sherratt, A 1991 'Sacred and Profane Substances: The ritual use of narcotics in Later Neolithic Europe.' in Garwood P, Jennings D, Skeates R and Toms J (eds) *Sacred and Profane, Proceedings of a conference on Archaeology, Ritual and Religion* Oxford University Committee for Archaeology Monograph No 32, 51-64.

Sherratt, A 1995 'Alcohol and its alternatives: symbol and substance in pre-industrial cultures.' In Goodman, J et al *Consuming Habits: Drugs in History and Anthropology*. Routledge, London.

Sherratt, A 1996 'Flying up with the souls of the dead' *British Archaeology* June, No 15, 14.

Simpson, D 1996 'Excavation of a kerbed funerary monument at Stoneyfield, Raigmore, Inverness, Scotland (1972-1973).' *Proceedings of the society of Antiquaries of Scotland* Vol 126, 53-87.

Singer, C, Holmyard, E J, and Hall, A R (eds) 1979 *A History of Technology*. Volume 1 Oxford University Press.

Stewart, M 1985 'Excavations at Moncrieffe, Perthshire' *Proceedings of the Society of Antiquaries of Scotland* Vol 115, 125-151.

Stika, H-P 1996 'Traces of a possible Celtic brewery in Eberdingen-Hochdorf, Kreis Ludwigsburg, SW Germany.' *Vegetation History & Archaeobotany* Vol 5, No1-2, 57-65.

Stika, H-P 1996 'Keltisches Bier aus Hochdorf.' In Biel, von J *Experiment Hochdorf*, 64-76, Stuttgart.

Souden, D 1997 *Stonehenge*. English Heritage.

Soudsky, B and Pavlu, I 1972 'The Linear Pottery Culture Settlement patterns of central Europe.' In Ucko, P J, Tringham, R and Dimbleby (eds) man, settlement and Urbanism, London, Duckworth, 317-28.

Tannahill, R 1973 *Food in History*. London, Eyre Methuen.

Tilley, C 1996 *An Ethnography of the Neolithic: early prehistoric societies in southern Scandinavia*. Cambridge University Press.

Thomas, J 1988 'Neolithic Explanations revisited: the M-N

transition in Britain and S. Scandinavia.' *Proceedings of the Prehistoric Society* 54, 59-66.

Thomas, J 1991 *Rethinking the Neolithic*. Cambridge University Press.

Thomas, J 1996 'Neolithic houses in mainland Britain and Ireland, a sceptical view.' In Darvill & Thomas *Neolithic Houses in Northwest Europe and Beyond*. Neolithic Studies Group Seminar Papers 1, Oxbow Monograph 57, 1-13.

Thorpe, I 1996 *The Origins of Agriculture in Europe*. Routledge, London.

Todorova, H et al 1983 'Ovchorovo. Raskopi I Proucavania 8.' *Archaeological Institute of the Bulgarian Academy of Sciences.* Sofia (referred to in Bailey 1996).

Traill, W and Kirkness, W 1936 'Hower, a prehistoric structure on Papa Westray.' *Proceedings of the Society of Antiquaries of Scotland* Vol LXIII, 309-323.

Troels-Smith, J 1967 'The Ertebølle culture and its Background.' *Palaeohistoria* 12, 505-28.

Tyndall, J 1876 Addressing the Glasgow Science Lectures Association, quoted in Conant 1954.

Vencl, S 1994 'The Archaeology of Thirst' *Journal of European Archaeology,* 2.2, 299-326.

Wainwright, G and Longworth, I 1971 'Durrington Walls Excavations 1966-1968.' Society of Antiquaries of London.

Weissner, P and Schiefenhovel, W (eds) 1996 *Food and the status quest: an interdisciplinary perspective.* Berghan Books.

Whittle, A 1985 *Neolithic Europe*. Cambridge University Press.

Whittle, A 1996 'Houses in Context: Buildings as process.' In Darvill T and Thomas J (eds) *Neolithic Houses in northwest Europe and Beyond*. Oxbow Monographs 57, 13-27.

Wickham-Jones, C R 1990 *Mesolithic and later sites at Kinloch Bay, Rhum*. Society of Antiquities of Scotland Monograph Series No 7, Edinburgh.

Woolf, A & Eldridge, R 1994 'Sharing a Drink with Marcel Mauss: The uses and abuses of alcohol in early Medieval Europe.' *Journal of European Archaeology* 2:2, 327-340.

Zohary, D and Hopf, M 1993 *Domestication of Plants in the Old World: the origin and spread of cultivated plants in West Asia, Europe, and the Nile Valley.* Clarendon Press, Oxford Science Publications.